Life in the Oceans

World of Wildlife:
Life in the Oceans

ORBIS PUBLISHING · LONDON

From the original text by Dr Félix Rodríguez de la Fuente
Scientific staff: P. de Andres, J. Castroviejo, M. Delibes, C. Morillo, C. G. Vallecillo
English language version by John Gilbert
Consultant editor: Dr Maurice Burton

© 1970 Salvat S. A. de Ediciones, Pamplona   © 1975 Orbis Publishing Limited, London WC2
Printed in Italy by IGDA, Novara   Reprinted 1978   ISBN 0 85613 499 6

Contents

Chapter 1: The ocean: cradle of life   2
Chapter 2: The seashore: frontier of two worlds   27
Chapter 3: Sea turtles and sea snakes   51
Chapter 4: The birds of the coastal fringes   69
Chapter 5: Mammals of the seashore   101
Chapter 6: The gorgeous world of the coral reef   125
Chapter 7: The frozen wastes of Antarctica   139
   Order: Sphenisciformes
Chapter 8: Life on the Continental Shelf   157
Chapter 9: The fishes of the shallow seas   167
   Class: Pisces
Chapter 10: Cormorants, frigate birds and boobies   199
Chapter 11: The intelligent, friendly dolphins   217
Chapter 12: The wide oceans and the abyss beneath   237
Chapter 13: Killers of the deep   249
Chapter 14: Wanderers of the ocean   259
Chapter 15: Birds of the open sea   269
Chapter 17: The mighty whales   279
Bibliography and Index   301

Acknowledgements

D. Anderson/NHPA: 78
A. Bannister/NHPA: 132, 161
L. E. Battaglia/Photo Researchers: 127
M. Beebe/Photo Researchers: 235
Ch R. Belinky/Photo Researchers: 59
H. Beste/Ardea Photographics: 2
Betzeler/Bavaria: 140
L. C. Bissel/Nancy Palmer: 208, 268, 270, 271
S. C. Bisserot/Bruce Coleman: 214
J. Blossom/NHPA: 114
M. Brosselin/Jacana: 72, 87
J. Burton/Bruce Coleman: 34, 39, 40, 45, 50, 57, 61, 72, 75, 218
Miguel Ángel L. Castaños: 41
Cedri: 34
Ernesto Cerra: 53, 73, 80, 84, 98, 105, 107, 163, 169, 176, 197, 202, 222, 238, 250, 252, 255, 265, 274, 282, 293
R. Church/F. P. G.: 219, 227, 232, 246, 247
Bruce Coleman: 60
R. Church/Photo Researchers: 3, 170
S. Dalton/NHPA: 71, 86, 94, 96
R. Dawson/Bruce Coleman: 79
J. P. Dupont/Jacana: 262
F. Erize/Bruce Coleman: 87, 88, 198, 206
F. Erize: 207, 209
Antonio Escudero: 10, 13, 18, 30, 49, 141, 162, 164
Faulkner: 37, 132, 137
A. Giddings/Bruce Coleman: 118
S. Gillstater/Tiofoto: 68
E. W. Grave/Photo Researchers: 240
F. Greenaway/NHPA: 95
A. Greham/Photo Researchers: 10, 13, 18, 30
A. Gutiérrez: 264
J. Hardening/Black Star: 257
J. G. Harmelin: 35, 156, 160, 163, 185, 188
M. P. Harris/Bruce Coleman: 122, 208
G. Holton/Photo Researchers: 142, 273
E. Hosking: 81, 85, 93
D. Hughes/Bruce Coleman: 55, 60
R. Jacques/Photo Researchers: 113
P. Johnson/NHPA: 117, 138, 153, 154, 276
R. Kinne/Photo Researchers: 9, 21, 175, 183, 225, 228, 278
G. Laycock/Bruce Coleman: 104, 197, 211
José Lalanda: 15, 16, 31, 32, 38, 45, 70, 77, 82, 88, 89, 92, 93, 104, 106, 114, 116, 118, 128, 130, 132, 133, 136, 146, 149, 168, 185, 187, 193, 194, 201, 205, 207, 211, 214, 215, 224, 229, 234, 239, 256, 260, 271, 272, 275, 276, 277, 280, 285, 287, 294, 296

Ch Lederer/Bavaria: 124
L. Lee/Bruce Coleman: 211
A. Margiocco: 24, 25, 241
Marka: 99, 144, 145, 235
G. Martens/Bavaria: 40
Mateau Sancho: 108, 150, 151
G. Mazza: 33, 37, 39, 43, 46, 48
S. McGutcheon/F. W. Lane: 100
T. McHugh/Photo Researchers: 105, 171, 185, 196, 216, 244, 253, 263, 299
D. Middleton/Bruce Coleman: 76, 200
G. J. H. Moon/F. W. Lane: 90, 91
G. Musseley/Jacana: 190
K. B. Newman/NHPA: 204, 213
NHPA: 275
Noailles/Jacana: 261
F. Pellegrini: 121, 122
P. Morris/Ardea Photographics: 195
A. Power/Bruce Coleman: 26, 42, 47, 66, 135, 183
E. Puigdengolas: 44, 179, 191, 221, 258, 261, 266
F. Quilici/Moana: 5, 234
C. Ray/Photo Researchers: 20, 300
P. Richter/Ardea Photographics: 97
P. Reisere/Bavaria: 164, 173, 179, 183, 192
F. M. Robert/C. E. Ostman: 36, 40, 240
J. G. Ross/Photo Researchers: 231
J. Rychetnik/Photo Researchers: 110, 111
F. Sauver/Bavaria: 194
R. Schroedar/Bruce Coleman: 54, 56
V. Serventy/Bruce Coleman: 228
F. Shulke/Black Star: 248, 254, 257, 284, 287
J. Simon/Bruce Coleman: 22
J. Simon/C. E. Ostman: 299
J. Simon/Photo Researchers: 59
J. Stevens: 289, 295
Marcelo Socias: 6, 17, 102, 196, 226
Suinot/Jacana: 143
V. Taylor/Ardea Photographics: 65, 129, 134, 135, 166, 187
E. Twelves/NHPA: 143
J. P. Varin/Jacana: 115
J. Vasserot: 46
P. Ward/Popperfoto: 63
J. Wightman/Ardea Photographics: 74
D. P. Wilson: 12, 19, 29, 159, 165
Zentrale Farbbild Agentur GmbH: 11, 35, 136, 164, 192, 290
Zimmerman/Alpha: 177

Foreword

Last great unexplored region on our planet, the ocean, occupying more than two-thirds of the earth's surface, has long been a realm of marvel and mystery. Reputedly populated by fabulous and terrifying monsters, the sea, where life began, is now beginning to yield up its secrets. This tenth volume of *World of Wildlife* concentrates on the astonishing animal communities of the 'silent world', from the sands and rocks of our familiar coasts to the darkest reaches of the abyss, more than six miles below the surface.

The complexity of life in the ocean is suggested by the visible forms to be found on the seashore – shells, limpets, barnacles, crabs, starfish, sea anemones and hosts of other tiny animals whose lives are regulated by the ebb and flow of the tides. Fishing in the shallow, well-lit waters where green plants grow and where the major part of marine life is concentrated are sea birds magnificently adapted for gliding, diving or swimming; and in tropical zones the rainbow-hued inhabitants of the coral reef obey the implacable laws of nature, engaged in a deadly, never-ending game of hide-and-seek.

Contrary to earlier beliefs, the physical features of the ocean bed are now known to be as varied as any equivalent region on land. In a seascape of immense ridges, lofty peaks, caves, chasms and trenches, amazing fishes roam the perpetually dark waters of the deep. Some look like floating skeletons, some have massive heads and murderous teeth, others literally light up in the black abysmal depths.

Equally remarkable are the animals that roam freely over the high seas, including the sharks, largest of all fishes, long-distance flying birds such as petrels and albatrosses and the large mammals which have adapted to life in a marine environment. Having solved the vital problems of breathing, feeding and keeping their temperature steady, seals and sealions visit land only to breed; and the intelligent dolphins and whales spend their entire lives in the ocean.

World of Wildlife

Life in the Oceans

CHAPTER 1

The ocean: cradle of life

The shifting frontier between land and sea is the home of many animals, some terrestrial, some marine, others leading an amphibious existence.

For centuries man believed that the earth was flat, like a gigantic plate, and that one huge river, its bounds stretching to the horizon, snaked around its circumference. For primitive man, and indeed right down to the Middle Ages, this seemingly endless belt of water was a hostile element, inhabited by monsters – a realm of terrifying legend and superstition. The destiny of man, who was, according to Biblical tradition, fashioned by God from clay, undoubtedly lay on land, on what he termed 'earth'. It was only later, as a result of the experiences of daring sea voyagers, that astronomers and cartographers came to the inescapable conclusion that the earth was round and that, furthermore, the larger part of it was covered by ocean.

Although there is evidence of long-distance sea travel centuries before Columbus and Magellan, it was only then, and in the following decades, that western man began to appreciate the sheer immensity of the seas surrounding his familiar coasts. As explorers sailed out to discover new lands it became possible to gauge the area and volume of the oceans in relation to land masses; but it was to be some time before man came to understand the true significance of the ocean as literally being the cradle of life itself.

A new and revolutionary perspective of our planet has recently been revealed, giving us an even clearer conception of our modest place in the cosmic scheme. As filmed by astronauts in outer space, the earth is shown to be a blue planet, surrounded by a bright aquamarine halo, glistening like a tiny jewel in an infinity of blackness. And this is what distinguishes it from other planets and stars millions of miles away. Our blue oceans, our humid atmosphere, our biosphere – that part of the earth where living organisms exist—are in all probability unique in the solar system. It can all be seen clearly from space—the seas, the clouds that move across the continents, and the falling water which helps to sustain land plants and animals. Today there can be no room for doubt that our earth is an aquatic planet and that all the living organisms to be found on it derive from the ocean depths.

Man returns to the sea

Every summer holiday-makers in their hordes invade the beaches, drawn irresistibly, it would seem, to the sea. On these visits, man, if he is sufficiently inquisitive and perceptive, can discover phenomena which are now accepted as fundamental to the fairly new science of oceanography. The amateur fisherman, for example, will quickly realise that certain fish species are invariably to be found in particular zones, as in coastal waters. The birdwatcher will be aware that the birds which perch on cliff ledges and swoop low over the seashore are almost always the same—gulls, terns, plovers and sandpipers—and that puffins, razorbills and the like congregate on lonely offshore rocks. The sporting fisherman, with the facilities to venture out onto the high seas will have more spectacular species at his disposal, such as tuna, bonito and sailfishes; and the birds following his

Certain groups of mammals, having spent a period on dry land, eventually returned to the sea. These animals, known as cetaceans, include the dolphins, one of which, the bottle-nosed dolphin, is illustrated here.

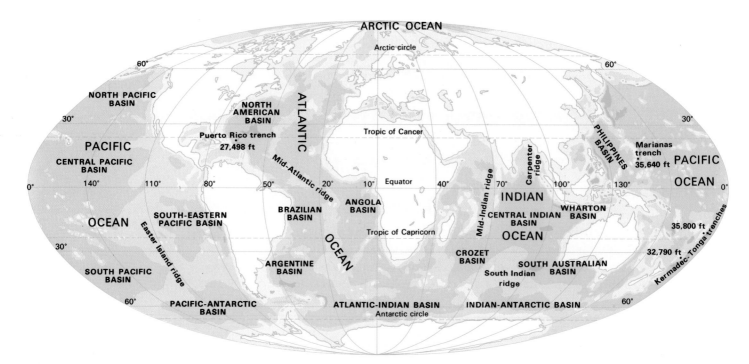

It was once believed that the sea bed was completely flat. It is now known that it is made up, in many areas, of deep trenches and high ridges. The average and maximum depths of the ocean are greater than the corresponding heights on dry land.

vessel may include shearwaters and, in tropical seas, pelicans. A passenger on an ocean liner may catch sight of stranger fishes and, in some latitudes, will perhaps glimpse a whale; while overhead soar petrels and albatrosses, daring sea birds which can survive so far from land by finding ample food beneath the waves. By now the traveller has progressed well beyond the coastal waters which cover the continental shelf. Below him, thousands of fathoms down, stretches the abyss.

Basically this is the pattern which we shall adopt for describing life in the various parts of the ocean. Admittedly, it is somewhat arbitrary, for many sea animals pass freely from one zone to another. But since some method has to be devised, it makes sense to discuss, in turn, the animal communities of the coastal fringes, then those of the continental shelf, and finally those to be found in what is loosely termed the open ocean or high sea.

If, instead of restricting himself to surface fishing, our imaginary seafarer decides to go in for underwater exploration, he will immediately notice that in the areas closest to shore there is enough sunlight to distinguish the colours of fishes and other marine animals. The experienced skin-diver, with suitable equipment, can venture deeper to the point where, although animal life is plentiful, the water is not so well illuminated. Farther than that he cannot go. At the edge of the continental shelf, where it dips sharply downward, the water becomes so dark and the pressure so enormous that not only he, but also most of the sea creatures he has so far encountered, come up against an impenetrable barrier.

In many parts of the world the three characteristic zones—coastal belt, continental shelf and abyssal depths—are clearly defined, each with separate and distinctive animal communities. But this pattern is by no means general, for sometimes the abyssal plain slopes very gradually, without any transitional stage, from the shore; and elsewhere a ship may steam hundreds of miles away from the mainland without leaving the clear, comparatively shallow, continental waters.

Our survey of life in the ocean will take us, therefore, from the coasts to the high seas, and from the surface down to the depths. The animal communities of the beaches and cliffs, some linked closely to the land, some to the sea, others amphibious, are extremely varied, including mammals, birds, reptiles, fishes and invertebrates. In shallow waters—the only places where tiny algae and plankton can exist—there is a complex pattern of life where animals born on land, such as mammals and birds, mingle with fishes and certain invertebrates which never leave the water. But in the intermediate zones and, above all, in the black depths of the abyss, only organisms that are highly specialised can survive. These are the realms of phosphorescent fishes, of giant squids and of those champions of underwater diving—the whales.

How big is the sea?

If we were to imagine the earth as reduced to the size and shape of an orange, the ocean would be represented as a thin film of water. On a cosmic scale, therefore, the sea—regarded as an inexhaustible source of food and a convenient outlet for all waste matter—seems relatively insignificant. But in terms of the biosphere, that small portion of the earth where plants and animals are capable of surviving, the ocean assumes enormous importance. For, in the whole solar system, only the earth is believed to be abundantly supplied with water.

It would appear that in the history of our planet seas and oceans have never occupied an area larger than they do today. They cover approximately 140 million square miles, roughly seven-tenths of the earth's surface. Their average depth is around 13,000 feet, the deepest trenches extending to almost seven miles (the maximum sounding being 35,640 feet in the Marianas Trench, south of Guam). It is worth recording that the average height of the land areas is 2,800 feet and that the highest mountain, Everest, is only 29,028 feet. It has been calculated that if all the continents were submerged, filling the oceanic basins, the depth of remaining water would still average 8,000 feet above the level of the sunken land masses.

Origins of the oceans

Where did the sea come from? What were the origins of those great expanses of salt water? At one time it was believed that the ocean basins were flooded by prolonged periods of torrential rain, brought about by the large quantities of condensing water vapour in the primitive atmosphere. The supposition was that enormous storm clouds unloaded rain on the still white-hot earth, causing immediate evaporation and the formation of new clouds. The gradual freezing of the planet would have created conditions in which some water would have been retained in liquid form, eventually giving rise to the seas and oceans as we know them today. Curious as it may seem, the principal

Cradle of all living forms, the sea contains a multitude of animal species. Doubtless many more remain to be discovered.

Water is an indispensable element of life and represents a high percentage of the body weight of all animals. In its three states, solid, liquid and gaseous, it helps to regulate climate and to maintain the physical equilibrium of the natural world. All water lost by evaporation is restored in the form of rain or snow.

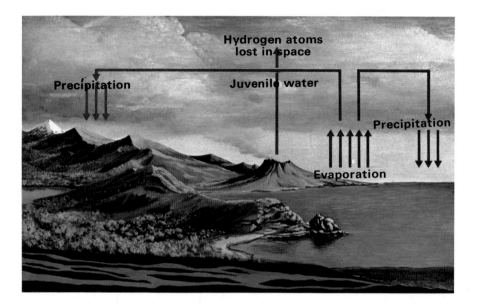

objection to such a theory (according to F. D. Ommanney) is that not enough water exists on our planet for it to have been derived from primordial rainstorms of this magnitude.

In an influential paper published in 1951, Rubey put forward an alternative theory, suggesting that the oceans, in the course of geological ages, certainly increased in depth even if they did not become larger, and that they, like the atmosphere itself, emerged from the earth's interior, being formed by chemical processes originating in the rocks making up the planet. The expanses of salt water, as they exist today, would therefore not have come about as a result of condensation of atmospheric vapour, but as a consequence of volcanic activity. This theory would partially explain the high content of dissolved salts found in sea water.

Whatever its origin, water is the essential element of the terrestrial biosphere, where it exists in three forms—solid, liquid and gaseous—assuming indispensable functions such as the regulation of climate and the maintenance of all living organisms. Water lost by evaporation is restored in the form of rain. In the oceans more water is lost than is fed back by rainfall, which is the reverse of what happens on land. It is possible that the most stable types of water, although doubtless subject to certain modifications, are those found in the oceans, in frozen form on the ice-caps, and in subterranean pools in the phreatic zone below the water-table.

Oceans and climate

In physics specific heat is the term applied to the capacity of a substance to absorb or release heat. The unit of measure is the calorie—namely the amount of heat necessary to raise the temperature of one gramme of water by one degree Centigrade. Considering that almost all substances have a specific heat lower than that of water, they reheat more easily when subjected to irradiation and likewise freeze more rapidly. Thus, during the day, the earth's crust, warmed by the rays of the sun, absorbs heat as quickly as it loses it during the night. But in the ocean the fluctuations and contrasts of day and night temperatures are far less striking. Because of this the sea helps to regulate the surrounding air temperature. If oceans did not exist, the earth would be a furnace by day and a mass of ice by night, making life impossible. Proof of the effect of sea on climate is the fact that the hottest and coldest places on earth are in areas far removed from the ocean.

Water, in addition, possesses another physical property which plays some part in regulating the earth's temperature — latent heat — a term used to describe the calorific changes which occur when a substance passes from one physical state to another. When one gramme of water in liquid state is converted into ice, it releases 80 calories; the same quantity of heat is absorbed by this weight of water when it is transformed from a solid to a liquid state. In evaporating, one gramme of water absorbs 539 calories, which it releases again as it condenses.

Similarly, the ocean releases heat when it freezes in the polar regions and absorbs it when evaporation occurs in the tropics, thereby affecting the temperatures of both areas. Henderson has calculated that at the equator approximately 7.5 cubic feet of water to every square foot of surface evaporate each year, implying an absorptive capacity of billions of calories per square mile.

The latent heat of most substances is, like the specific heat, lower than that of water; consequently water is the only element in the solar system capable of performing a temperature-regulating function. It is significant, therefore, that water, exuded by sweat glands or by panting, helps to regulate the body temperature of homoeotherms or warm-blooded animals, and that is equally important, although in a different way, for poikilotherms or so-called cold-blooded animals, whose body temperatures fluctuate according to outside conditions. Because water accounts for approximately nine-tenths of the composition of every living organism, there is no risk of damage as a result of extremes and sudden changes of body temperature.

Submarine mountains and canyons

The sea covers not only the ocean basins and depressions but also areas of continental land, the submerged fringes of which form what is commonly known as a continental shelf.

This formation was until quite recently considered to be a fairly level, gently sloping slab of land, as on the central Atlantic coast where the change of level is indeed barely 5 inches per mile, and elsewhere even less. It was assumed to extend to a depth of about 600-650 feet. We now know that in certain regions the continental shelf terminates only 300 feet from the surface (as in the Mississippi delta), but that in other places it continues down to as much as 2,300 feet, as off the shores of Florida. Nor is it surprising to learn that it may vary greatly in dimensions. According to Carmina Vigili, it may extend, in some areas — off the coasts of Newfoundland, for example — some 550 miles out to sea. Yet around Africa it is hardly in evidence at all. In some cases the shelf may form the bed of immense 'eponti-

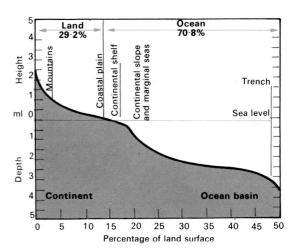

Cross-section of the earth's crust, showing the percentage of area occupied by continental masses and oceans (according to Sverdrup, Johnson and Fleming).

nental' seas, examples of which are the Philippine Sea and the Sea of Japan.

Oceanographers were for many years puzzled by the presence under the sea of V-shaped gorges or canyons, cutting deeply into the surface and edges of the continental shelf. They appeared to be continuations of valleys hewn out on dry land by fast-flowing rivers. According to M. J. Dunbar, however, this was unlikely because these rocky canyons evidently became progressively wider as they plunged below the 9,000-foot mark. Indeed, some of them are gigantic as, for example, that of Monterey, off the California coast, which in size is comparable to the celebrated Grand Canyon.

When, in 1936, Reginald Daly explained his theory about the origin of these underwater formations to the Geological Society of America, he began by saying that what he was about to announce was so extraordinary that he himself found it difficult to believe. But although nobody knew exactly how they were formed, he asserted that submarine canyons definitely existed, and suggested that these deep gorges were furrowed out by currents of muddy water, in conjunction with climatic factors (such as the melting of glaciers), which flowed continuously and rapidly down the submarine slopes. Although there is supporting evidence for this theory, notably the undoubted presence of so-called turbidity currents, there are arguments against it, principally the thickness and hardness of eroded rock. Despite the valuable work of oceanographers and geologists, we still do not know precisely how such canyons originated.

The continental slope—the point at which the slight incline of the shelf gives way to a sharp, precipitous drop—was once thought to be the border of the ocean basins; but recent in-

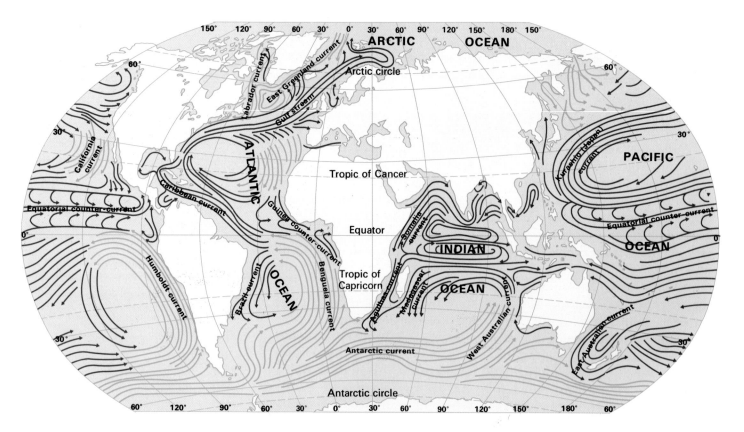

Ocean currents are determined by variations of water density, by winds and, in part, by tides. They have a profound influence on the climate of the sea and shores of the lands around which they flow. This map shows some of the principal currents of the world's major oceans.

vestigations have revealed that there is a transitional zone, or precontinental fringe. Topographically this differs little from the true ocean bed, but shows structural variations.

Scientists aboard the first ships used for oceanography, equipped with ballasted ropes and cables, reached the conclusion that the ocean bed was an enormous flat plain, hardly sloping, and broken here and there only by volcanic formations which occasionally projected above the surface to form oceanic islands. It is now known, however, that far from being a simple abyssal plain, the ocean bed is as varied, in its relief, as the surface of land continents. Apart from flat areas there are underwater formations of mountain chains (one of the best known being the mid-Atlantic ridge) as well as isolated peaks and deep trenches, the characteristics of which are not fully understood but which are generally thought to be of a volcanic nature. As a rule, slopes of such trenches contain very little, if any, sediment. In this respect they differ from other submarine formations.

In coastal regions, movements of the water often give rise to ripples in the mud or sand. In some places far from the shore fossil ripples indicate that in the distant past such areas were covered by the ocean.

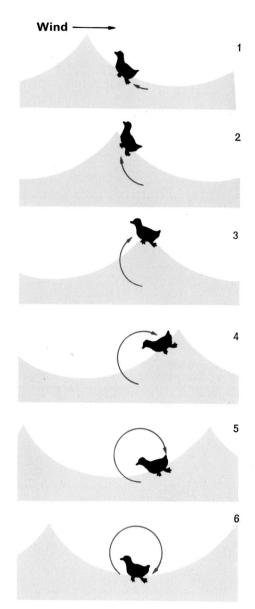

Waves are caused by wind but the movements of the water are vertical, not horizontal. This can be proved by placing an object in the sea. No matter how strong the waves, this will bob up and down but will come to rest in the trough of each successive wave, having moved little if any farther forward.

Different theories have been advanced to explain the origins of ocean basins. According to one, they were formed by a succession of sinking and rising movements of continental masses; according to another, continents which were linked in primeval times later split and drifted apart, rather like icebergs. The originator of this theory of continental drift, in the early 1900s, was Alfred Wegener, but not until 1950 or thereabouts was it seriously considered by scientists attempting to prove that continents had shifted in relation to the magnetic poles. Today most geologists accept the principle of Wegener's theory. They have studied the various phases of the process and believe that the submarine ridges and peaks are part and parcel of continental land masses, from which, at some distant time, they came to be separated.

Biology of the ocean

Light, salinity, temperature and pressure are the main variable factors characterising the ocean, where, broadly speaking, there is more than sufficient oxygen for the needs of its inhabitants. Since only plants are capable of using solar energy for the synthesis of organic matter, all the living creatures in the sea depend on vegetation for nourishment. The plants, in their turn, cannot survive unless exposed to sunlight. About 99 per cent of the ocean is plunged into total and perpetual darkness. A large amount of sunlight is already dissipated at the surface by reflection and the rest is absorbed only a few feet below the water. Light, of course, is not really white, but is made up of seven colours forming the visible spectrum, to which may be added infra-red and ultra-violet rays. In general the red is very quickly absorbed by the water; but the blue rays, also used in the process of photosynthesis, penetrate much deeper. In the Sargasso Sea, for example, where the water is particularly transparent, they reach a depth of about 500 feet before being reduced to one per cent of their power. In coastal waters, according to Gunnar Thorson, daylight is reduced to one-hundredth of its strength between the depths of 30 and 100 feet. But the use of sensitive instruments has, according to the same author, shown sunlight still having an effect at 2,600 feet below the Mediterranean and at over 3,000 feet in the Caribbean. The level at which there is sufficient sunlight for plants to develop is called the photic zone.

In spite of its high specific heat, the surface temperature of the ocean varies greatly according to latitude. In the polar regions the water temperature hovers throughout the year around 0°C (32°F); but in the Persian Gulf the annual mean temperature is around 24°C (75°F). Above a depth of 600 feet or so the temperature remains almost constant all the year round. Down between 6,000 and 10,000 feet it is equally stable, approximately 4°C (39°F), whether it be at the poles or at the equator or at any point between these.

One of the most notable characteristics of the sea is its salinity, which is usually expressed in terms of a thousand parts of water. More than three-quarters of the dissolved salts in the ocean consists of sodium chloride, common salt. The rest is

made up, in decreasing amounts, of magnesium chloride, magnesium sulphate, calcium sulphate, calcium carbonate and potassium sulphate. Apart from these, other minerals, including gold, silver and zinc, have been traced, but in such minute quantities that they can only be expressed in terms of million parts of sea water. Some secondary elements, such as copper, cobalt and vanadium have also been identified, because certain animals feed on them selectively.

The salt content of sea water fluctuates only very narrowly from place to place. Near the poles, where evaporation is minimal and a large amount of liquefying ice turns to fresh water, salinity is approximately 34 parts per thousand. The most extreme limits are in the Red Sea, where salinity is 43 parts per thousand, and in the Baltic, where it is only 7 parts per thousand It has, however, been determined that the proportions of dissolved salts in sea water, whether high or low, remain constant,

Because of the various movements of the sea—waves, tides and currents—the composition of its water remains virtually constant.

doubtless as a result of being stirred about by the ocean, always in movement.

It is surprising that similar, sometimes identical types of animals, such as certain sea cucumbers, marine worms and molluscs, live both on the continental shelf and on the ocean bed, at more than 32,000 feet. This was demonstrated in 1951 by the scientists of the Danish ship *Galathea*. Gunnar Thorson believes that pressure has no damaging effects on animals which lack internal air-cells, and that such creatures can move about equally freely, without special adaptations, either on the continental slopes or in the abyss. It has been estimated that pressure increases by about one atmosphere (the standard unit of air pressure at sea level—about 14.72 lb per square inch) for every 30 feet of depth.

The restless sea

The composition of sea water is fairly uniform, due to the fact that mixing processes are continually at work. At the upper level, at any rate, the ocean is for ever moving, under the influence of tides, waves and currents.

All the heavenly bodies attract one another. The earth is principally affected, by reason of proximity, by the moon, and of course by the sun. Because the earth's crust is rigid, the effects of sun and moon on land can only be detected with the aid of special instruments; but in the sea things are different. The pull of sun and moon, combined with the rotational movement of the earth upon its axis, are responsible for the tides, the rhythms of which are enormously important for people living in coastal regions and whose livelihoods often depend on the sea. Every twenty-four hours there are two high tides and two low tides, at roughly six-hour intervals. Every fifteen days the pull of the sun, reinforcing that of the moon, produces a particularly strong tide, with maximum rise and fall—the spring tide. Midway between these there is a weak tide, with minimal fluctuations—the neap tide—caused as the earth forms a right angle in relation to sun and moon.

The most obvious movements of the sea, however, are brought about by surface waves driven by the winds. This produces a continual undulating motion or swell, which is simply an up-and-down movement with hardly any perceptible forward progression. The phenomenon is proved by the fact that an object floating on a wave-tossed sea bobs up and down but does not move along with the crests. The greater the distance between the wave and the one following, the higher it will rise without breaking. Occasionally, waves of this kind may build up far out to sea and travel thousands of miles, eventually crashing against the shore and perhaps doing heavy damage.

The most unpredictable, terrifying and potentially destructive movements of the ocean are caused neither by planets nor winds but are the consequence of gigantic underwater earthquakes and volcanic eruptions. The waves formed by these submarine upheavals may sweep across the sea at incredible speeds, sometimes at 400–500 miles per hour, their crests 75–100 feet high, venting

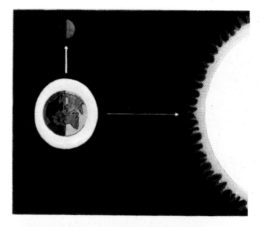

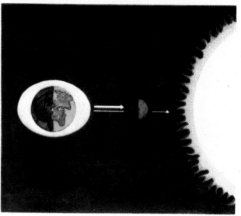

Tides are produced by the pulls on the earth of the sun and moon. When they are at right angles to the earth (*above*) they cancel out and cause weak or neap tides; but when sun and moon are in line, their combined pulls bring about strong spring tides (*below*).

Facing page: Long regarded as a hopeful source of additional food for an overpopulated planet, the ocean's plant resources are now known to be limited, and because of widespread pollution the seas are in danger of being poisoned.

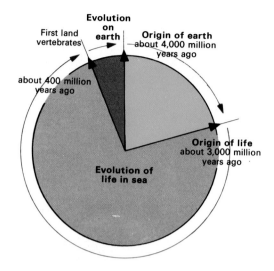

In this chronological diagram, the various sectors show the period during which no life existed on earth and the respective periods of evolution of living forms on land and in the sea. It is evident that much the greater part of the life-developing phenomenon occurred in the ocean.

their fury on some distant low-lying stretch of coastline or narrow, land-locked bay, carrying everything—trees, ships and buildings—in their wake. In Japan, which has probably suffered more than any other country from their devastating power, these tidal waves are known as tsunamis. In 1896 more than 26,000 people were killed by one such wave.

Wind, tides and differences of density (due to fluctuations of water temperature at varying latitudes) combine to produce ocean currents. It has long been realised that swift-flowing currents, comparable to underwater rivers and often clearly visible near the surface, have an important bearing on plant and animal life in the sea. They also exert a considerable influence on the climate of neighbouring lands.

The direction of the major ocean currents is determined, above all, by the rotation of the earth. In the northern hemisphere they tend to flow clockwise and in the southern hemisphere anti-clockwise. The zones where two currents meet are particularly rich in marine fauna. This arises from the fact that the confluence of currents brings about a mingling of life-giving mineral elements which may be absent in one or the other when flowing separately.

In contrast to the warm currents near the surface, there are cold currents, emanating from the polar regions, which, because of their greater density, flow deeper down. They are known as thermohaline currents, inasmuch as they are determined by the temperature and salinity of the water, although the physical structure of the ocean depths certainly has an additional influence on their movements.

Due, in large measure, to the different types of water movement, the oceans have for millions of years retained certain characteristics which have permitted life, in many forms, to develop. But today, for the first time, the oceans are in danger of becoming contaminated. The poisons and industrial wastes which man is heedlessly pouring into the sea may not only cause the local destruction of animal communities—serious enough—but could even provoke a situation whereby the natural purifying and restorative processes of the ocean would no longer operate.

The tree of life

There are approximately one million animal species on earth and of those more than 80 per cent are to be found on dry land. Nevertheless, the vast majority of these terrestrial forms are insects which, since they appeared, have multiplied and diversified to become the largest animal group. Setting insects aside, about 65 per cent of animal species have their home in the sea.

If, furthermore, we go back into the past, we find that all living creatures, including those best adapted to life on land, have had marine ancestors. Palaeontologists have established that all land animals had their origins in groups that were once confined to the sea. There came a time in the distant past when these animals abandoned the ocean and launched themselves into a new element, gradually adapting to an existence on dry land. This is why the tree of life may be said to have had its roots

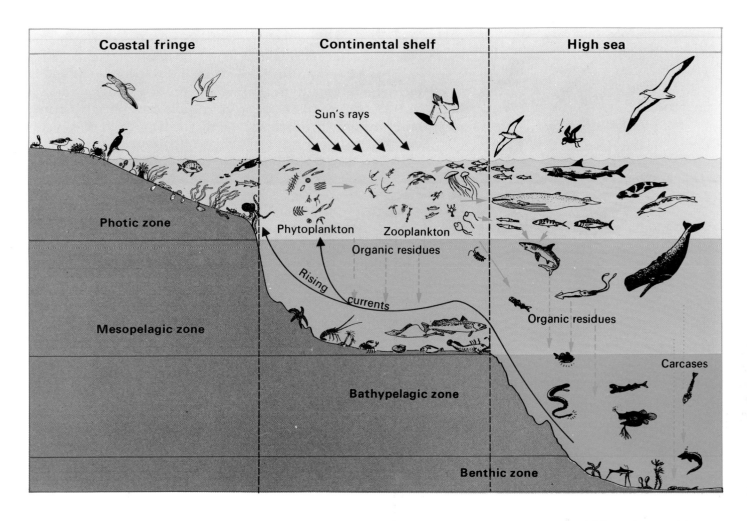

in the ocean, for it is there that the first living creatures were formed.

How, when and where did this miracle occur? What circumstances determined it and what forces contributed to the process? How is it possible that from an extremely simple structure (as was certainly the case with the earliest living forms) an enormous diversity of plants and animals developed, ranging from trees and bacteria to whales and man himself? In attempts to answer these perplexing and challenging questions, scientists have put forward a number of conflicting theories.

In the light of most recent investigations, it is estimated that the earth was formed approximately four and a half thousand million years ago and that the first living creatures appeared some one and a half thousand million years later. During the long period when the planet was dead the conditions which eventually enabled life to emerge were slowly shaping and intermingling. It is thought that these phenomena occurred in the sea, under circumstances completely different from those that exist today, involving a synthesis of large organic molecules, some of which acquired (how and when is not known) the capacity of reproducing themselves. Laboratory experiments have proved that in an atmosphere comparable to that enveloping the earth in primeval times, an electric discharge is capable of producing certain molecules of amino-acids, essential components of living matter.

The first living creatures on earth were not yet capable of

From the coastal fringe, battered by waves, to the depths of the ocean known as the abyss, there are many complex plant and animal communities. The surface zones, illuminated by the sun, are inhabited by a multitude of species. The intermediate zones are populated by many organisms which depend for food on debris drifting down from above. In the abyss there is an accumulation of mineral elements, vital for the development of algae, which, because of rising currents, are borne up towards the surface.

manufacturing organic matter from mineral salts and water by means of solar energy, in the manner of modern plants. This complex process, photosynthesis, was only rendered possible at a later stage of evolution. Before the miracle of recognisable life occurred, the food of these primitive creatures consisted of organic molecules which swarmed so densely in the water that they formed a kind of highly nutritious 'soup'. Since Oparin first advanced the theory in 1936, biologists have come to agree that such organic compounds must have preceded the genesis of living creatures, determining, by natural selection, their subsequent development.

Little by little, on a temporal scale in which human life can be measured as a millionth of a second, living forms multiplied, diversified and evolved in the ocean. The conquest of dry land only began some 350 or 400 million years ago. How did it come about? It is believed that representatives of different groups of animals, in complete independence of one another, broke through the barrier separating the water from the land. The successful colonisers engendered new forms of life which, in turn, expanded and branched out, taking possession of the many ecological niches offered by their new environment. Nevertheless, the majority of species remained for ever in the sea. In that environment animal communities are subjected to conditions very different from those that exist on land, but the basic essentials for survival are the same—to eat and not to be eaten, to reproduce and to leave the largest number of descendants. These vital necessities exercise the identical selective pressures in the ocean depths as they do on the savannah or in the rain forest.

From surface to abyss

Up to a few years ago it was assumed that the amount of vegetable matter produced by the sea was much greater than that of the land. Recent calculations relating to the possibility of the oceans providing supplementary food for man have indicated, however, that marine plants are far less abundant than had been hoped; and since, additionally, the food chains in the sea are far more complex than on land, there is little chance of the oceans furnishing an inexhaustible supply of food for a steadily growing human population.

It has been estimated that for every million photons (quantities of light energy) striking the surface of the earth, only 90 are used by plants for manufacturing organic matter. Of these, about 50 will be utilised by land plants and the remainder by marine plants. In the sea the density of the water soon acts as a barrier to the penetration of sunlight, so that the processes of photosynthesis, origin of every food chain, can only take place in the surface zones. It is true that there are in the ocean bacteria capable of producing organic matter without the aid of light, due exclusively to chemical reactions, but their influence is minimal.

The concentration of plants at upper levels determines the pattern of ocean life. It is in this surface zone that plant-eating

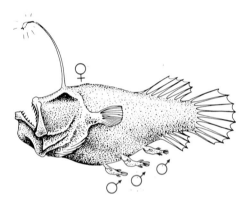

Among pelagic fishes, such as herrings, which live in shoals, reproduction is usually effected by means of eggs, laid in large numbers by the females and then fertilised by the males. In the depths, however, there are various alternative methods. One of the most elaborate is that of the angler fishes. The dwarf males responsible for fertilisation spend almost their entire life attached as parasites to the bodies of the larger females.

Facing page: Basically the food chains of the ocean, though more complex, follow the same pattern as those on dry land. At the base of the ecological pyramid are tiny plants—phytoplankton—which use solar energy to create organic substances from water and mineral elements. Higher up the scale come consumers, phytophages and predators.

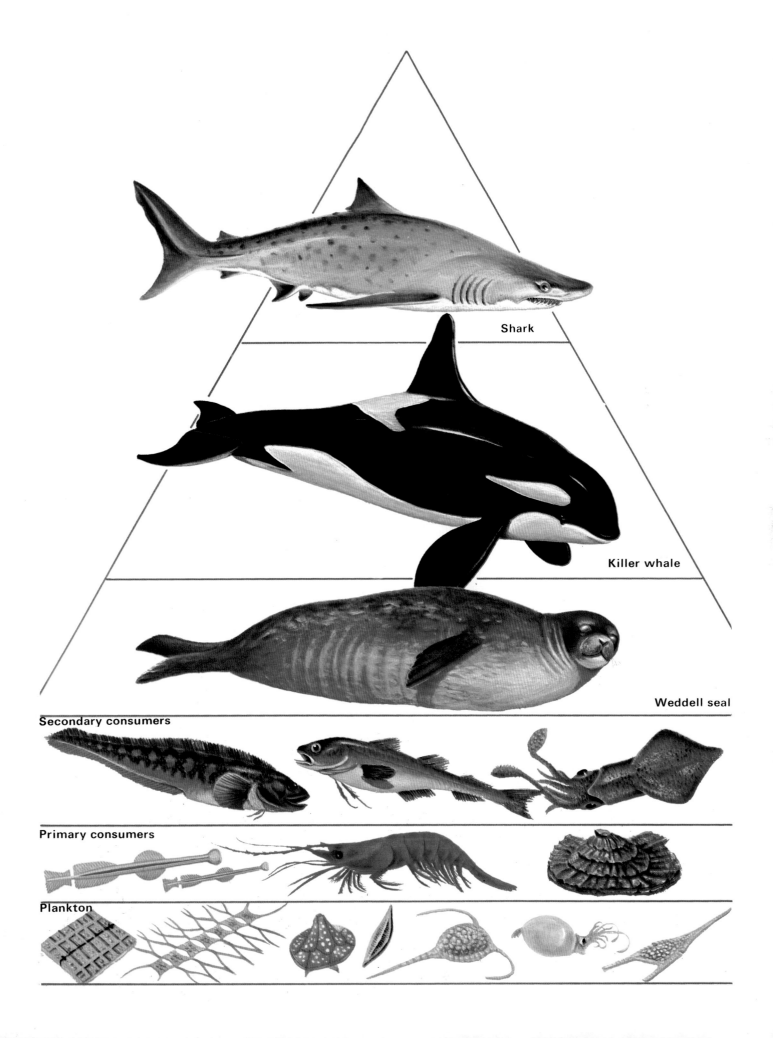

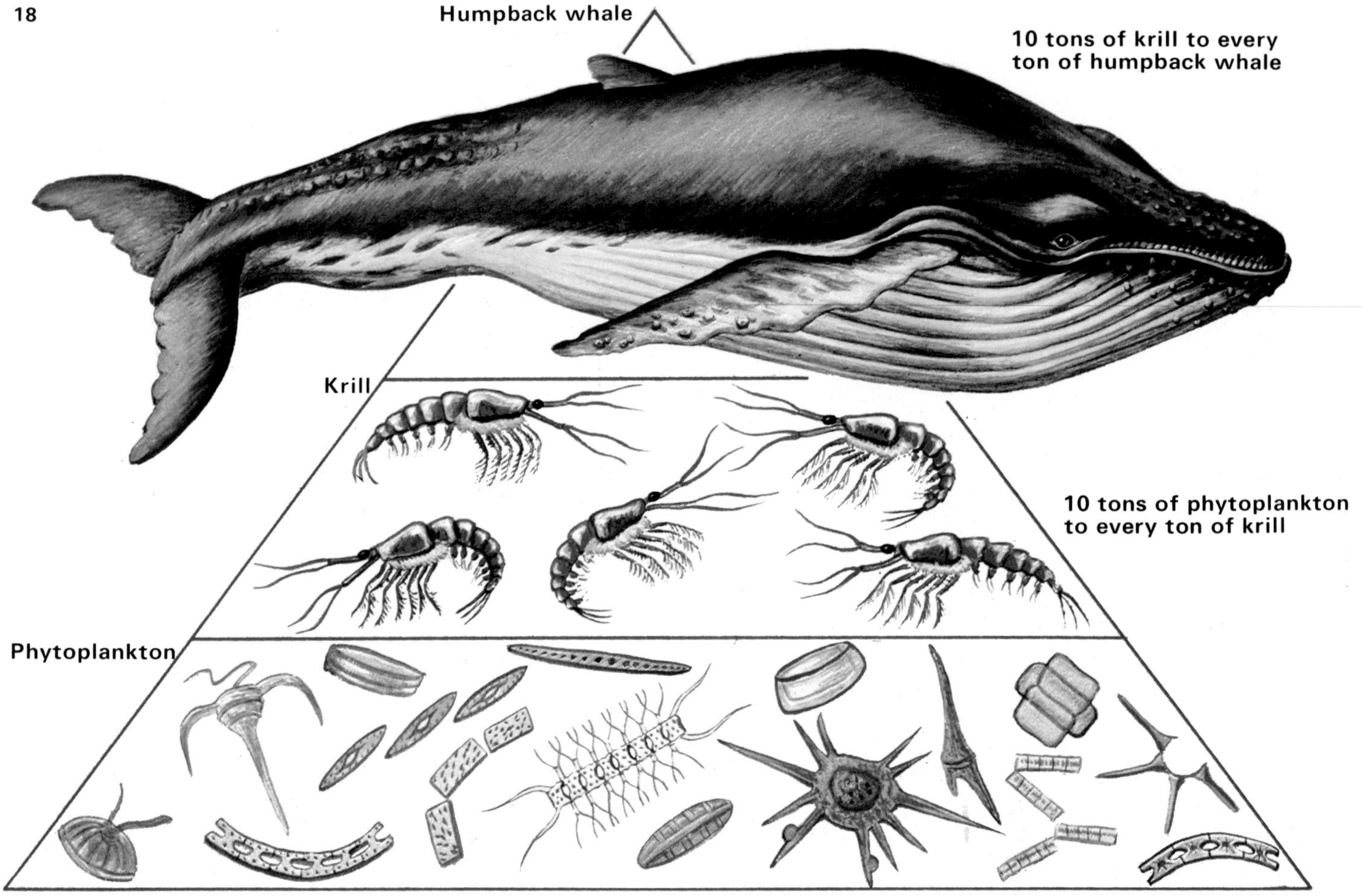

Humpback whale

10 tons of krill to every ton of humpback whale

Krill

10 tons of phytoplankton to every ton of krill

Phytoplankton

animals are found which, in turn, attract carnivores. But sooner or later all creatures living near the surface die, and their bodies fall to the ocean bed. In the course of the journey to the abyss they undergo decomposition by bacterial action. The mineral components collect in the depths but because of the darkness cannot be used again by plants. But in the intermediate zones between surface and abyss the continual 'rain' of organic matter provides food for species adapted to life in semi-darkness. Such species, of course, only retain these substances briefly, until they too die.

In the depths conditions are very different from those near the surface or in the intermediate zone. Pitch darkness prevents the growth of plants capable of photosynthesis, but the accumulation of minerals enables a complex community to flourish. The animals of the deep include filter-feeding organisms which have no need to move about, because currents continually provide them with water that is rich in food. They therefore have a solid substratum where they can crawl, burrow or attach themselves, invaluable for escaping enemies.

These remarkable plants and animals of the ocean bed or abyss are scientifically known as benthic communities, and are to be found in greatest density and variety in comparatively shallow seas where the minerals collecting at the surface, and circulating fairly freely in the lighted zone, lead to the appearance in the upper levels of the ocean of a rich flora and fauna. But only in areas where ascending currents bring up minerals that have collected on the ocean bed towards the surface are conditions suitable for the development of the microscopically small algae which constitute the first link in the various food chains of the open sea.

Baleen whales feed directly on zooplankton – in their case tiny shrimps known as krill. They therefore occupy a relatively low level of the ocean's food pyramid.

Facing page: Were it not for phytoplankton, microscopic algae drifting in the photic zone of the sea (*below*), there could be no life in the ocean. Living on these plants are zooplankton (*above*), made up of tiny crustaceans and of eggs and larvae of marine animals, also found close to the surface.

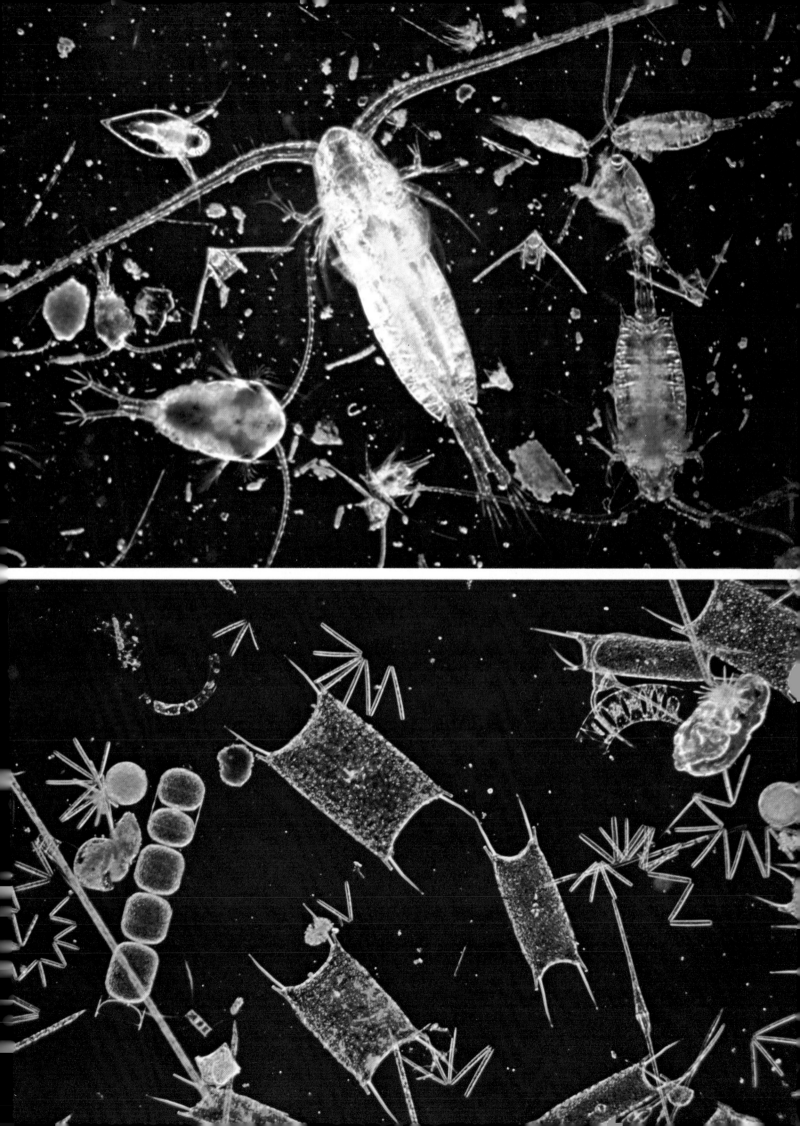

Related to corals, but capable of moving freely in the water, the jellyfish possesses tentacles with stinging-cells, so that contact with the animal may cause painful injury.

The drifting food of the ocean

If the layman were asked to describe typical algae he would probably think immediately of the branched or ribbon-like seaweed commonly seen close to the shore, either stranded at low tide or thrown up on the beach by the incoming tide. Certainly this description holds good for many algae which are freely visible, but not for the vast majority of forms which remain permanently concealed below the surface of the sea. The mass of ocean flora is in fact constituted of simple, single-celled organisms which are much too small to be seen by the naked eye, except on the rare occasions when certain species form huge floating clumps on the surface.

Suspended in the water, where they find the substances necessary for their survival, the microscopically minute algae are forever drifting to and fro, carried at random by the ocean currents. Because of the weak penetrating power of the sun's rays they can only develop at or close to the surface, and are best likened to an immense floating prairie, continually in movement. Mingling with these tiny plants, and equally at the mercy of wind and wave because they lack any means of independent locomotion, are innumerable marine animals, just as small, together with the eggs and larvae of a large number of other ocean species. These organisms, drifting passively in rhythm with the currents and tides, are known collectively as plankton; but the plant particles are specifically described as phytoplankton and the tiny animals as zooplankton. There is a further distinction among the latter group, namely holoplankton (animalcules which spend their entire life as zooplankton) and meroplankton – creatures which only appear in the guise of plankton for a certain part of their life cycle, as, for example, the eggs and larvae of marine animals which eventually depart to live on the sea-bed.

The composition of plankton may vary in the same area, according to time and season, but there is a broad distinction between neritic (or coastal) plankton and oceanic plankton. The richest and most varied growths are to be found near the shore where there are a large number of habitats. On the high seas the pattern is more uniform. Thus river estuaries, beaches, rocks, cliffs, coastal lagoons and the like harbour a wide range of phytoplankton, an additional development factor being the shallowness of the water in such zones, which facilitates the upward movement of minerals to the surface. Furthermore, just as there is a change of pattern away from the shore, there are fluctuations downward as well, with a thinning out of species and quantities with increasing depth.

Finally, although some species are found all around the world, there is a notable difference in plankton composition according to latitude. The Arctic plankton consists of relatively few species and its principal constituents are diatoms (in the case of plants) and, among zooplankton, copepods. In temperate, tropical and subtropical seas, on the other hand, there are many more species. But in the Antarctic it consists mainly of krill, tiny shrimps that are eaten by some whales and other animals.

The concentration of zooplankton in any given area and at any given depth varies by day and by night. As the sun gradually sinks to the horizon its rays obviously form ever-narrowing angles with the surface of the sea and their powers of penetration become progressively weaker. As soon as night starts to fall innumerable planktonic organisms begin slowly ascending to the upper levels and when dawn breaks they once more descend. The causes of these daily up-and-down movements of zooplankton are not fully understood but it is more than probable that differences of temperature are largely responsible for the phenomenon. Certain organisms spend the day in deep water which, because it is colder than the water above, enables them to keep their body temperature constant. There is also little doubt that light is an important consideration. It has been proved, for example, that the mere passage of a cloud across the surface is enough to provoke a general upward movement. This suggests that the same process must occur when zooplankton are found in a zone where phytoplankton is abundant, the density of the latter to some extent screening the sun's rays. Nevertheless, we do not know precisely why such organisms shun the light. Perhaps the tiny animals in seeking the darkness of the muddy depths simply find protection from predators, making for the surface where food is plentiful only when increasing shade and gloom make it less likely that they will be detected.

Whatever the real reasons for these daily shifts towards the surface and back to the depths, it is clear that dispersal is much

The open sea offers no natural shelters, so many plant-eating fishes form large shoals for self-protection. The effect may be either to lead a predator astray or to deceive it into mistaking the shoal for a single, enormous fish.

The largest ocean animals are mammals—the cetaceans, of which the most massive is the blue whale, over 100 feet long and weighing up to 130 tons. The pilot whale, illustrated here, is small by comparison. An adult may measure up to 28 feet, which is about the size of a blue whale at birth.

facilitated by the continuous activities of waves, tides and currents. The zooplankton community is enormously varied. Many of the constituent organisms are plant-eaters but there are also predatory forms as well as some which feed on detritus. Although the majority of these drifting organisms are microscopic (as, for example, bacteria and flagellates which measure less than 5 microns) or very small indeed, there are several species which attain considerable dimensions—namely certain forms of jellyfish, in particular the Portuguese man-of-war (*Physalia physalis*) and common sea blubber (*Cyanea capillata*), which may be more than 3 feet in diameter.

Life near the surface

Because plant growth is virtually restricted to the surface layers of the ocean, this is where the majority of animals are to be found. It is a community consisting of species which have adapted to a life of incessant swimming.

In the ocean, as on land, life is a constant struggle for survival. The constituent elements of the food chain are identical—producers (photosynthetic plants), consumers (plant-eaters, carnivores, scavengers) and decomposers. Here, as everywhere, the principle is 'eat in order not to be eaten', and the laws of natural selection favour those species best equipped either to capture prey or to defend themselves.

Participation in this ferocious drama of life and death has brought about remarkable adaptations, both anatomical and physiological, in the case of many marine animals. The streamlined shape of tunas, the extraordinary scenting powers of sharks,

With rational exploitation, the sea could be a rich source of food for humanity. The past activities of the whaling industry are examples of how to squander such resources. It has led to the massacre and near-extinction of certain species. As a result of international legislation, whereby some species are not allowed to be hunted and the females and young of others are similarly protected, it is to be hoped that numbers will be stabilised.

the sonar mechanisms of dolphins and the group-hunting techniques of killer whales are only a few random examples. But the serious problem for species that are hunted is that the ocean offers no shelter or hiding places. Consequently, many surface fishes have adopted what might appear to be a self-defeating procedure—that of forming immense shoals. In fact this type of group behaviour has a double advantage in that it confuses the enemy, confronted as it is with a cloud of fishes swimming in all directions, and can even prove deceptive, since such a moving mass may take on the appearance of a single adversary much larger and stronger than the predator itself.

Another means of escaping predators is sheer size, especially, for example, in the case of the baleen or whalebone whales, which are not only larger than any other living mammals but even more immense than the extinct dinosaurs. These giants of the ocean feed on plankton, and were it not for man they would be virtually invulnerable.

The intermediate zone

Below the illuminated surface zone and the depths of the sea is a vast expanse of water which is either in partial shadow or in perpetual darkness. The scarcity of light hinders the development of phytoplankton and the absence of a solid base or substratum prevents the accumulation of residues floating down from the surface. The animal inhabitants of this intermediate zone have no alternative but to trap such particles as they drift down, before they sink to the more inaccessible levels of the abyss. Under such conditions animal populations are not dense,

Many predatory fishes of the abyss have adapted to the conditions of their environment in astonishing ways. Some are equipped with relatively large jaws which can be opened so wide as to engulf victims even bigger than themselves.

though in places sufficiently large to attract sperm whales and other large predators. Indeed, certain species are only known to science because their remains have been found in the stomachs of dead cetaceans.

Shortage of food at the greater depths has resulted in certain deep-sea predators acquiring the most astonishing hunting techniques. Roaming the abyss, for example, are fishes with enormous jaws, some of which are capable of swallowing in a single gulp prey considerably larger than themselves. Here too are numerous species with luminous organs that serve one or more purposes. They may be defensive, designed to break up and disguise the animal's shape, or they may be offensive, as when situated at the end of a long stalk protruding from the mouth or head, when they serve as a lure for prey. Some abyssal crustaceans even envelop themselves in a phosphorescent cloud for purposes of camouflage, comparable to the inky clouds emitted by squids when they sense danger. Alternatively, the light signals may be used among different members of a species for mutual recognition in the breeding season.

Meeting for the purposes of reproduction is a major problem for species in an environment where individual numbers may be small and where it is always dark. 'Lighting up' is one way of overcoming this difficulty but there are other, equally ingenious methods. Probably the most perfect example of adaptation is that of the angler fishes. The males, very small and in some cases dwarf-like, literally become parasites of the larger female, attaching themselves to her body and surviving by sharing her blood circulation system. The only duty of the males is fertilisation, so that they simply develop into repositories of sperm, without any independent existence.

The ocean bed

Until the middle of the 19th century it was assumed that beyond a depth of 1,700–2,000 feet there was no life in the ocean. But when in 1860 the underwater telegraph cable linking Sardinia and Tunisia was raised from a depth of over 6,000 feet it was found to be encrusted with marine animals. This discovery aroused widespread interest in oceanography and a number of expeditions were organised, the most spectacular of which was the voyage of H.M.S. *Challenger* (1872–76), exploring the Atlantic and Pacific Oceans. In 1951 scientists on board the Danish ship *Galathea* brought to light living animals from the Philippines Trench, 34,600 feet down; and in 1960 Donald Welsh and Jacques Piccard descended in a bathyscaphe to a depth of 35,800 feet. Since then there have been a number of other attempts to probe the mysteries of the deep, among them the dives of Commander Jacques Cousteau and his colleagues in *Precontinent* I, II and III, and the American 'Sealab' experiments.

The ocean bed is the only place in the sea where marine organisms can attach themselves to a solid substratum. Its diverse structure encourages the formation of different types of animal community, while its irregularity creates certain zones where currents flow strongly and others where they can hardly be detected. Lack of light prevents the growth of plants, but all the residues from the upper levels of the ocean settle here so that it is unnecessary for energy to be wasted by swimming around continuously in quest of food. Yet not all the inhabitants of the deep feed on this detritus. The larvae of many species, which spend the early parts of their lives as constituents of plankton, are undoubtedly a rich source of food as they drift towards the bottom. Furthermore, photographs taken at considerable depths with cameras to which bait is attached reveal the presence of many predators unknown to science.

A final mention must be made here of certain organisms that by the very specialised nature of their development create a ready-made environment for many fascinating marine species. These are the corals of tropical seas, which form beautiful islands and reefs where a wide range of fishes and other ocean animals flourish.

Although some fishes of the deep are blind, others have enormous eyes that enable them to reap the best advantage of what little light exists. These grotesque individuals are a species of hatchet fish.

CHAPTER 2

The seashore: frontier of two worlds

Although it is on the seacoast that the destructive powers of the mighty ocean are all too often in evidence, it is here that one also sees it most clearly as a positive, creative force. The small pools that fill rock crevices after storms and high tides harbour an extraordinary wealth of marine life. Beaches, lagoons and low shorelines are the habitats of countless wonderful plant and animal communities. We may be accustomed to thinking that the most spectacular forms of coastal life are the many sea birds—gulls, terns, plovers, sandpipers and the like; but the truth is that there is an equally fascinating, more secret, world at our fingertips—the muddy realm of worms, crabs, snails and other tiny creatures, which provide food for the shoals of fishes teeming in the offshore waters.

Conditions of life on the seashore are, in some ways, as rigorous as those confronting the inhabitants of the abyss or, on land, the residents of the desert. But survival is no simple matter in a zone where the essential element—water—comes and goes at regular intervals. Aquatic animals living on beaches, subjected to the rhythmic ebb and flow of the tides, have the advantage over land forms when the tide is high; but at low tide they are faced with the problem not only of continuing to breathe but also of escaping an army of predators patiently waiting for receding water to expose sand and shingle.

Even more of a threat to such animals than the periodic movements of the tides are storms which may whip the waves into a frenzy. When tons of water hurl themselves against the cliffs and eat away the shoreline, a multitude of tiny organisms would be in danger of being crushed, were it not for special forms of adaptation, whether physical, in the guise of a protective shell, or behavioural, as in the instinct to burrow. In some cases the

Facing page : The seacoast, dividing land from ocean, is the home of many zoological groups, many feeding on plankton, many carnivorous. One of the fiercest of the marine predators is the moray eel, lying in ambush among rocks and coral.

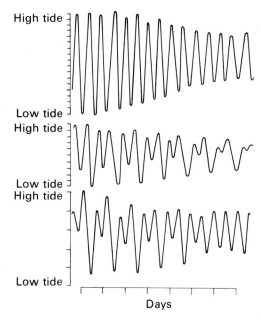

The rhythm of spring and neap tides is partly dependent upon geographical location. In the case of semi-diurnal tides (*top*) the extension of two successive tides is identical. Where diurnal tides (*centre*) occur, there may be smaller, sometimes imperceptible, fluctuations, between recognisable high and low tides. Finally there are mixed tides (*bottom*) which combine both the afore-mentioned types.

Facing page: Many areas of coastline assume completely different guises at high and low tide. The area in between the high and low water marks is known as the intertidal zone. It contains many fascinating animals which have to adapt to continually fluctuating conditions.

unusual anatomical structure of the organism concerned will determine the reflex action which enables it to withstand the fury of the elements and avoid being dashed to pieces.

Gazing down at the sea from the top of a high cliff, one cannot fail to be impressed by its sheer grandeur. Tragic and terrifying when huge waves batter the rocks below, serene and welcoming when sunlight glitters on the waves, streaking a golden path to the horizon, the ocean is a symbol of mystery and eternity. But although it may arouse aesthetic instincts and poetic images in us, it has a much more basic, brutal impact on the innumerable inhabitants of the seacoasts. For them it can be a paradise one moment, a hell the next, all depending on the whims of the weather.

The coastal scene

The many thousands of miles of coastline bordering the continents and islands of our planet assume diverse forms, some of which have hardly been studied, so that it is impossible to compress them all into a single description. Nevertheless, whether considering tropical shores or icy polar coasts, there are certain features that remain constant, making it feasible to summarise the coastal scene and the principal factors influencing the development of its flora and fauna.

The seashore may be formed either of rocks, with scattered boulders, stones and pebbles and often fringed by steep cliffs, or of mud and sand. These two principal types of landscape owe their respective origins to the slope of the continental shelf, for this influences the movements of the water and determines the geological structure of neighbouring areas of exposed or periodically exposed land, of which it is simply a continuation. Under the constant impact of the waves, the hard substratum is gradually eroded. Little by little, chips and blocks of rock are broken off, either being swept away by the tides or remaining in place, there to be slowly worn down by wind and wave. These inorganic materials, reinforced by those derived from the disintegration of the skeletons of marine animals, are transformed into sand or mud, depending on the size of the grains. Carried to and fro by the sea, they are deposited in areas where the lie of the land is suitable for the accumulation of sediment, and in this way beaches of sand or mud are formed.

When the solid substratum of the continent comes into contact with sea water, it is influenced by several other factors as well. It is of course subjected, to a lesser or greater degree, to humidity. The amount of moisture is minimal in zones merely splashed by waves and affected only by spray, but greatest, naturally, in areas that are permanently under water. Furthermore, the salts dissolved in the sea or released in the atmosphere by spray furnish the soil with new chemical substances. The combined influence of these mechanical, physical and chemical phenomena transform the continental mass in such a way as to produce the characteristic shapes of coastline.

The tides, except in certain parts of the world as, for example, in the Baltic and the Mediterranean, play an essential role in

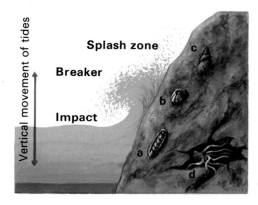

On rocky coasts the action of the waves delimits three distinct zones. The supralittoral zone is subject only to splash or spray at high tide. The eulittoral or intertidal zone is washed by incoming and outgoing breakers; and the sublittoral zone is the region directly subjected to the continuous battering of the waves, usually perpetually submerged. Animals living at these different shore levels survive in various ingenious ways. Chitons (a) cling tightly to the rock face. Acorn barnacles (b) can move their legs in and out, according to the state of the tides. Periwinkles (c) use what little humidity there is in the air to keep moist; and sea anemones (d) shelter in rock clefts.

determining the development of seashore life. The maximum level of water (high tide) and minimum level (low tide) encompass the so-called intertidal or littoral zone, where plants and animals alike are subjected to successive wet and dry periods. In addition to the rhythm of the tides, submerging and exposing this zone of beach four times daily, there is another cycle, based on the conjunctive pulls of sun and moon (responsible for all tides). The effect of this joint influence is greatest at the time of the spring tides and least at the time of the neap tides. The spring tides, with their extreme high and low levels, coincide more or less with the full and new moon. Midway between, at the first and last quarters, come the neap tides, when changes of level are less marked.

The movements of the tides assume different proportions, according to geographical location. On a world scale, three basic types may be distinguished. Semi-diurnal tides are characteristic of the Atlantic where two high and low tides a day reach similar levels. In the case of diurnal tides, as in the Gulf of Mexico, there appear to be only one high and low tide daily, but in fact the intervening tides are imperceptible. Finally there are mixed tides, common to the Indian and Pacific Oceans, in which both the previous forms are combined, so that the two high and low tides may be identical or uneven, depending on whether the semi-diurnal tide is predominant.

Waves—whose dimensions are linked with atmospheric factors and whose impact varies in relation to the slope of the continental shelf—have their effect on the ecosystem of the seashore. Some parts of the beach are washed by them four times a day as the tides come in and go out, and these alternating movements are responsible for marking out distinct zones, depending on how far and how high the waves reach. The lowest of these zones, directly bathed by the waves, not only receives the highest amount of humidity but is also subjected to the most powerful battering. It extends from the line of actual contact with the sea to the upper limit of the breakers on the shore. Beyond that is a zone where the waves, their strength sapped, merely lap the sand or shingle; and finally, higher up the beach, is the zone where the only sea moisture is provided by spray blown by the wind.

The degree of impact of the sea on the shore therefore gives rise, within a comparatively limited area, to a wide range of ecological niches, each of which is occupied by a community of plants and animals that are perfectly adapted to the particular conditions imposed upon them.

Other factors playing an important role in seashore life include light, oxygen and temperature. Sunlight, indispensable for the development of autotrophic plants, namely those that build their own nutritive substances, has much more of an impact here than in other parts of the sea, except on the surface. Coastal waters, therefore, are very rich in plankton and organic particles, which serve to feed large numbers of molluscs, crustaceans, worms and other interesting animals, these in turn attracting marine predators. Oxygen dissolved in water, the only form that can be utilised for breathing by sea animals,

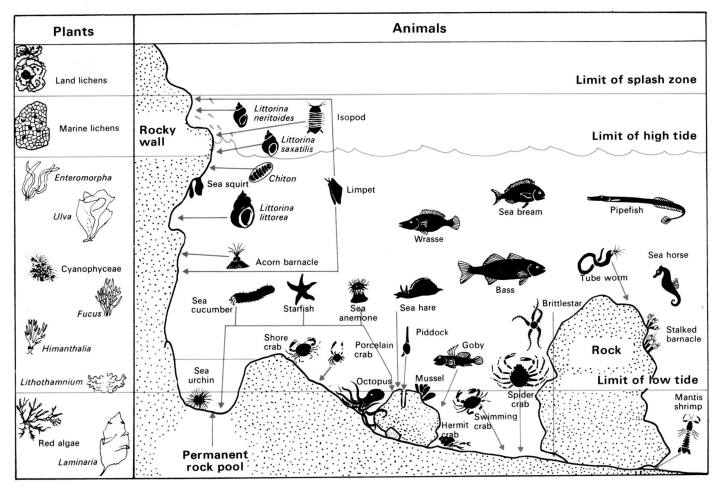

Distribution of certain typical plants and animals of rocky coasts.

is also more abundant in coastal waters than at other levels of the ocean, since the water, ceaselessly battering against the shore, is regularly and continuously aerated. Finally, the temperature of the seacoast is somewhat different from that existing inland, for the presence of this vast expanse of water attenuates extremes and establishes a comparatively stable climate.

It is very probable that conditions prevailing on seacoasts (exposed as they are to all these factors in conjunction) were, in times long past, responsible for the development of all living forms now populating our planet.

Mechanisms of survival

Although, broadly speaking, the biological conditions of coast and shore favour the proliferation of innumerable species, the rhythmic actions of the waves constitute serious obstacles to survival. In the intertidal zone particularly, plants and animals have been compelled to acquire adaptive mechanisms in order to overcome the problems of an environment which is alternately submerged and exposed. Thus certain species associate with others to form communities which help one another, while some will become parasites and live at the expense of their hosts. All manner of ingenious systems of attack and defence are brought into play. Individuals will hide, efface themselves against a surface so as not to be detected, resist water pressure by means of

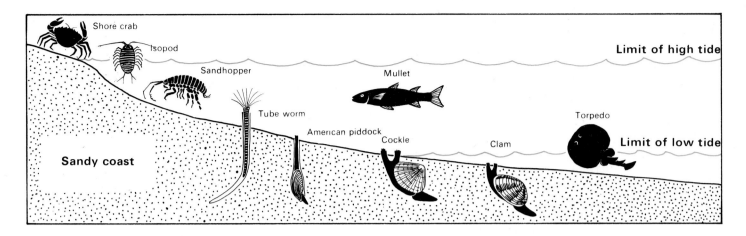

The distribution of animals along sandy shores and adjoining waters differs from that of rocky coastlines for two important reasons. The impact of the waves is far less violent and the substratum is soft and yielding. This diagram shows how some of the most characteristic species are distributed.

streamlined shapes, or transform themselves into self-sealing chambers in order to avoid drying out.

In bare places where there is no natural shelter the only method of self-defence against the continual, relentless battering of the waves is to remain solidly fixed to some type of support and expose the smallest possible surface. Certain animals have a remarkable body structure which enables them to fulfil the first imperative. The limpet is a mollusc whose shell is conical and flattened, and whose foot—a muscular extension of the body—adheres firmly to rock, the grip being so strong that it is impossible, except by prising it away with a knife, to tear the animal from its support. The sharp edges of the shell fit so tightly against the rock surface that the limpet is virtually enclosed in a sealed chamber where it can not only survive for long periods of drought but where it is practically invulnerable. It was long believed that the limpet clung to solid rock by a form of suction, but this is evidently not so, for the contact between animal and surface is so close as to leave not the smallest crack. Certain sea anemones stay firmly attached in much the same manner.

Equally ingenious is the procedure adopted by the acorn barnacles, crustaceans with slender cirri or bristly legs, which also have the appearance of truncated cones and which literally stick themselves to a rock surface by means of a viscous substance which hardens as time passes. This fluid is secreted in the carapace, consisting of tiny pieces of shell fitting neatly into one another. Once the acorn barnacles are submerged by the sea they stretch out their bristly legs, the rhythmic movements of which create currents that bring in floating particles. When the tide goes out, leaving them exposed, the animals close their shell, remaining moist and protected. As they shut themselves they let out tiny bubbles of air which produce a gentle crackling sound, quite audible to anyone attentive enough to the changes that are regularly produced on the seashore by animals and physical conditions alike as the tide ebbs.

Other animals attach themselves to solid surfaces by different methods. The attachment organ of the mussel, for example, known as the byssus, is made of bundles of strong threads. With the aid of this organ the mollusc not only clings firmly to the rock but also retains some freedom of movement. The bryozoans or sea mats, invertebrates that form colonies, are already protected against the onslaught of the waves by their minute size but in addition attach their exoskeleton to hard

Facing page: Sea anemones are carnivorous animals. Waving their tentacles gently to and fro in the water, they attract passing prey which are promptly poisoned by the stinging-cells and conveyed to the mouth by the tentacles. Most are fixed by the base more or less permanently. The *Cerianthus* shown here lacks a pedal disc and is a tube-dwelling burrower.

Organisms which possess no particular mechanisms for withstanding the pull of the outgoing tide and avoiding desiccation at such times, seek safety in sheltered spots. Sea cucumbers (*right*) live in rock clefts or in permanent pools. Certain crustaceans, such as the shore crab (*below*) make for beds of seaweed where conditions are always wet.

surfaces, forming a mat-like design which may also be seen on the shells of certain molluscs, on the fronds of algae and on the hulls of ships.

Sometimes organisms affix themselves only temporarily. Certain species that are normally very mobile only grip a solid surface when resisting pressure threatening to rip them away from such a support. This is the case with some cephalopods, whose tentacles are equipped with suckers.

Marine plants are likewise well adapted for clinging to surfaces. Among the large brown algae, for example, the lower extremity of the thallus or body ends in rhizoids—filaments which bear a superficial resemblance to roots but which, in fact, are quite different, serving merely as attachment organs. Other algae have a crust-like appearance and adhere to rocks along their entire surface, clinging to the tiniest fissures with such tenacity that it is virtually impossible to tear them loose without damaging them or the substratum.

The expanses of sand and mud which are continually exposed to the actions of the tides are the homes of many kinds of animals which lead a more or less sedentary existence. They are compelled, at regular intervals, to have recourse to one artifice or another in order to avoid the dangers to which they are subjected at low tide. Most of them protect themselves by burrowing in some fashion. Some dig tunnels or construct tubes into which

Coastal regions are the haunts of countless invertebrates which assume wonderful forms in adapting themselves to their surroundings. Tube worms (*left*) capture food by means of a tuft of outspread tentacles. Colonies of bryozoans or sea mats (*right*) may encrust solid objects in a delicate lacy pattern, or, as here, form lace-like fronds.

Stalked barnacles resist the force of the waves because they are attached to rock by means of a long, flexible, fleshy stalk, enabling them to move about with tides and currents.

they retire, blocking the entry. This type of behaviour is characteristic of the various species of bristle-worms, marine invertebrates which implant themselves in the mud. The only parts of the body to protrude are waving tufts of bristly tentacles which trap the tiny organisms on which the worms feed. Certain bivalve molluscs, such as clams, piddocks and other rockborers, bury themselves by burrowing into the sand, using a hatchet-like foot, or by boring into the rock. They remain in contact with the exterior by means of a siphon through which they feed, absorb oxygen and eliminate waste matter.

Certain toadfishes of tropical coasts (genus *Thalassophryne*) live perpetually buried in the sand, allowing their eyes, situated in the top of the head, to protrude. Others, such as representatives of the genus *Chaenopsis,* crawl into tubes left empty by marine worms. Blennies protect themselves at low tide by burrowing in the shingle where there is still enough water for them to survive.

Finally, a large number of seashore animals resolve such problems by avoiding the intertidal zone altogether and swimming out to greater depths. This happens among many fishes and crustaceans, notably shrimps. Some shrimps make for the high seas especially during the breeding season, for this guarantees a maximum chance of survival for their progeny during the period preceding the hatching of the larvae.

Cracks in rocks provide shelter for a horde of invertebrates which thus escape the violence of the waves. The very structure of these individuals indicates the relative security and calm of their surroundings. The limpets found here, for example, have taller, more pointed shells than the oval, flattened forms of those obliged to cling to rocks. These rock fissures are also favoured spots for sea anemones, starfishes, sea urchins, sea mats and different kinds of seaweed. These miniature aquariums are of endless interest to the naturalist and ecologist.

The rocky coastline

Rocky coasts and shores constitute a world of their own. Well lit by the sun, with plenty of oxygen and a solid substratum, these are places where a multitude of living organisms can flourish, even though, in certain parts, as already mentioned, the daily rhythm of the tides creates conditions where many individuals find it hard to survive.

Although no two stretches of rocky coastline are exactly alike, and the plants and animals to be found in surrounding waters will obviously differ according to geographical location, it is nevertheless possible to make certain generalisations, based on features which are constant, no matter where they occur.

The two main divisions of the marine environment are the benthic (sea bed) and the pelagic (all water above). Names for subdivisions vary but, broadly speaking, the benthic zone may be subdivided into littoral and deep-sea zones, and the pelagic into neritic (coastal, more or less coinciding with the continental shelf to where the water is about 600 feet deep), and oceanic, for all waters beyond this.

Although there is some disagreement among different authors, the littoral zone, which is the area from the high tide mark out to the limit of attached plants capable of photosynthesis, may be further subdivided into three zones, as already briefly described. The supralittoral zone is the area between the high water mark and the farthest limit of spray; the eulittoral (or intertidal) zone extends between the high and low water marks; and the sublittoral zone extends out and down to about 600 feet, the lower limit of photosynthetic plants.

The supralittoral zone is still really a part of the land environment for it is only submerged in exceptional circumstances. Vegetation consists of certain species of lichen and some halophilous plants, adapted to salt water. Apart from a few crabs there are no true marine animals here. Most of the insects, for example, are land forms.

The eulittoral zone, however, is teeming with marine life. In the upper part, which remains dry for long periods, one may find tiny crustaceans of the genera *Talitrus* and *Orchestia,* known as sandhoppers. Here too are periwinkles (genus *Littorina*) which feed on microscopic plants that cover rocks. On Atlantic coasts the species commonly found high on beaches is the small periwinkle (*Littorina neritoides*), which mingles, lower down, with populations of the rough periwinkle (*Littorina saxatilis*). These eventually yield place to the common or edible periwinkle (*Littorina littorea*). Such molluscs avoid desiccation by means of an operculum which blocks the opening of their shell, but the degree to which they are able to endure periods of exposure varies according to zonation. Thus *Littorina neritoides* can survive for as much as forty-two days in the open air, needing only a minute quantity of water to satisfy its functions. The common periwinkle can last for twenty-three days under similar conditions but is not often required to do so because its usual habitat is close to the low water mark. But the flat periwinkle (*Littorina obtusata*), the least resistant of the genus, can only endure lack of water for six days and is only found among dense colonies of algae where there is a particularly wet microclimate.

Rock cavities are the homes of many beautiful marine animals, such as starfishes (*centre*) and sea urchins (*left and right*).

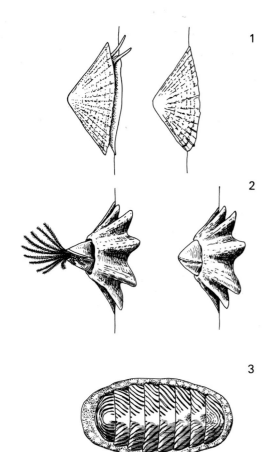

Among many small marine animals that attach themselves to rocks and other solid surfaces to withstand the force of the waves are limpets (*top*), acorn barnacles (*centre*) and chitons or coat-of-mail shells (*bottom*). Limpets affix themselves tightly by means of a fleshy foot. Acorn barnacles find refuge inside a shell of limy plates consisting of a series of plates through which the bristly thoracic legs protrude when conditions are favourable and which close hermetically when they are adverse. Chitons are attached along their entire under surface and their flattened shape offers little resistance to the waves.

Facing page: The three types of seashore animals described in the above diagram are seen here in characteristic protective attitudes. They are limpets (*above left*), chitons (*above right*) and acorn barnacles (*below*).

In the lower part of the eulittoral zone there are already many forms of algae, including the red algae of the genera *Corallina* and *Lithophyllum*. Then, in areas which are submerged for long periods, these mingle with a number of other forms (green, brown, red, etc), including the purplish-blue *Lithothaminion* species, the branching wracks of the genus *Fucus,* the knotted wrack (*Ascophyllum*) and the green sea lettuces of the genus *Ulva*. The lowest part of the shore, limit of the receding tide, is the almost exclusive domain of the oar weeds or kelps (genus *Laminaria*). Because the fronds of these algae retain so much moisture, they are the haunts of many animals—tiny snails, crustaceans, coelenterates, sponges, tunicates, worms and so forth, as well as innumerable larvae.

Other characteristic creatures of this intertidal zone include limpets, nut shells, chitons or coat-of-mail shells, sea hares, mussels and (both in rough and limpid water) acorn barnacles.

The echinoderms—sea urchins, starfishes, sandstars, brittle stars, feather stars, sea lilies, sea cucumbers and similar beautiful forms—are to be found all over the intertidal zone, particularly in sheltered hollows and pools. Typical of this large group are the various starfishes, such as the olive-green or brown (*Marthasterias glacialis*), and the tiny five-pointed star (*Asterina gibbosa*), and the red urchin (*Echinaster sepositus*). These starfishes surprisingly, are all carnivores, feeding principally on molluscs in a very unusual manner. Having used their arms to prise open the two halves of the shell of a mussel, for example, they then extrude their stomach through the mouth, enveloping and digesting the prey. Equally common in European coastal waters are the sea urchins of the genera *Paracentrotus* and *Psammechinus,* capable of inflicting painful stings, and the sea cucumbers, especially of the genus *Holothuria.*

The so-called sessile predators (attached to a base) are represented chiefly by cnidarians, armed with stinging-cells which paralyse or kill their victims. Among them certain sea anemones (genus *Actinia*) prefer spots that are almost always under water. Here too are many species of crustaceans with a somewhat irregular distribution, by reason of their mobility. These include the shore crab (*Carcinus maenas*), the pea crab (*Pinnotheres pisum*), the yellow crab (genus *Eriphia*) and many other arthropods. Especially interesting, on more than one count, is the hermit crab. Because the abdomen of this crab is soft and lacks a protective shell, it is a particularly tempting and appetising prey for a variety of seashore carnivores. The only means of defence for the hermit crab—its claws being of little use as weapons—is to scuttle into the empty shell of a gastropod, such as a snail or whelk.

Certain species of hermit crab enter into a symbiosis with sea anemones, a close association which proves to be of mutual advantage. The sea anemone profits from the fact that it can be transported and that it can consume any food left uneaten by the crab. The latter, in its turn, enjoys the protection provided by the polyp's stinging-cells, being immune to their poison, for these keep potential predators at a safe distance. Similar associations are to be found on coral reefs.

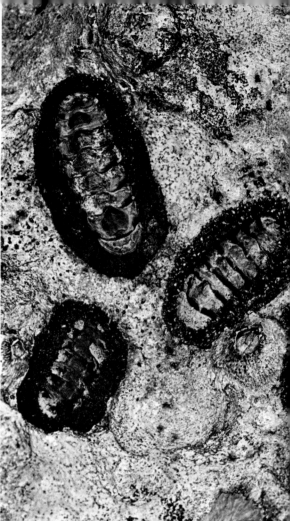

Clusters of polyps form virtual 'animal forests' in coastal waters. In zones that are permanently submerged beautiful, flower-like sea anemones flourish (*above*). Many marine animals fall prey to these carnivores, entrapped in their poisonous tentacles. Among them are blennies and several species of shrimp (*right*).

Fishes of the rocky coasts

Rocky seacoasts are the realm of benthos—plants and animals that either affix themselves to, or crawl along, the sea bed. The fluctuating rhythms of the ocean are incompatible with a pelagic life, although pelagic species do occasionally put in a fleeting appearance. Consequently, the characteristic vertebrates of the intertidal zone are almost exclusively benthic fishes. Birds and mammals frequent this zone principally with a view to procuring food, so that, to some extent, they tend to be strangers to the ecological community on which they exert such an influence.

The typical inhabitants of this region are a multitude of small organisms sharing what is a comparatively restricted space. Despite their modest size, all are perfectly adapted to their surroundings, demonstrating marked territorial habits and having recourse to complicated manoeuvres for the protection of their young—as dictated by the fact that they are at the mercy of many predators.

The most diversified family of fishes in coastal waters is that of the gobies (Gobiidae). Some 600 species have been classified, the common features being that none are more than 8 inches long and that they are customarily found either in pools formed by the outgoing tide, among the stones and pebbles of the sea bed, or in the 'prairies' of marine phanerogams far from the intertidal zone. For these fishes any cavity close to the shore (provided that it contains some water) will serve as a temporary refuge or breeding site. They may even shelter in empty tins, which clutter the beaches of so many holiday resorts.

Gobies have strong territorial instincts. The area within a radius of about 2 feet of the breeding site will be reserved by the male for rearing the young. He will defend it and get rid of any foreign bodies, live or inanimate, which may be present in the vicinity.

These shore and rock pools (as well as the empty shells of molluscs) are also the haunts of blennies (Blenniidae), distinguished from gobies by their more depressed shape and leaf-like ventral fins. Blennies swim close to the bottom and only leave their places of refuge to look for scraps of plants and detritus, which they consume together with large quantities of sand and tiny fragments of rock.

The clingfishes or suckers of the family Gobiesocidae also look much like gobies but instead of scales there is a thick layer of mucus covering and protecting the body, and the head is somewhat flattened. In adaption to being in constant contact with rock surfaces, the ventral fins have been transformed into suckers which are situated between the pectoral fins. It is these sucking pads which help to differentiate the species, the commonest Atlantic forms being the two-spotted clingfish (*Diplecogaster bimaculata*) and the shore clingfish (*Lepadogaster lepadogaster*).

Apart from the most formidable of all invertebrate predators, the octopus, many marine fishes with carnivorous habits have also adapted remarkably well to life in shallow waters, but are not so closely linked with rocky seacoasts as are the various

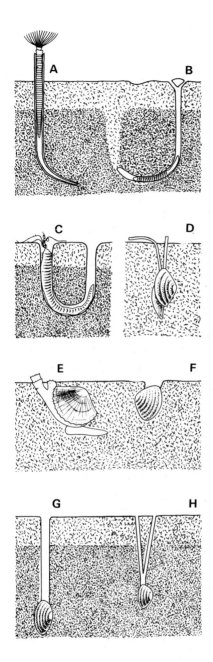

On sandy and muddy shores many animals burrow in the substratum, including the tube worms. Peacock worms (A) feed by coming to the surface and collecting floating organisms. Lugworms (B) swallow the mud or sand, and terebellids (C) consume detritus on the sea bed. Molluscs such as tellins (D), cockles (E and F), gaper shells (G) and furrow shells (H) bury themselves by means of their muscular foot. With the aid of their siphons they can draw in food particles and eject waste matter.

aforementioned species. All these hunters of small prey possess appropriate body structures with which they can dislodge their victims from the substratum. Thus the mullets of the family Mugilidae use their tough lips to gnaw at the surfaces of rocks, wrenching loose strips of seaweed as well as crustaceans and other soft-bodied organisms that happen to be seeking refuge there. They masticate their food partly by means of pharyngeal teeth situated on the last gill arches, and partly in the first compartment of the wall of the stomach which contains strong muscles and forms a crop. The sea bream of the genus *Diplodus* obtain their food in much the same manner, using their flat, sharp teeth to prise crustaceans and small molluscs from the rock face to which they are attached, grinding them into liquid form. The pipefishes (*Syngnathus*) and sea horses (*Hippocampus*) engulf prey by exerting powerful suction with their trunk-like mouth. The force of the sucking action is so strong that even a victim whose body is larger than the mouth opening of the predator concerned may be drawn towards it so violently that it will literally explode on contact and be easily digested.

The wrasses of the genera *Labrus* and *Crenilabrus*, usually found in zones where algae are abundant, are also predators, devouring seashore invertebrates in enormous quantities. Although of moderate dimensions, they are extremely aggressive. The only fishes capable of putting them to flight (in defence of their own territory or offspring) are bass and other species of comparable size.

The majority of large coastal predators are, however, shallow sea fishes which roam the shores because of the abundance of food to be found there. Moray eels (*Muraena*), with snake-like body and fearsome teeth, frequently conceal themselves in rock clefts just beyond the low-water mark. Certain elasmobranchs (the group of vertebrates including sharks and rays) also pursue their prey close to the shore. They include the torpedo or electric ray (*Torpedo marmorata*). The cells of its electric organs, situated on each wing and formed of modified muscles, are capable of giving out a shock of between 45 and 220 volts, enough to stun large victims.

Various flatfishes, such as the flounder (*Platichthys flesus*), also venture close to the beaches and are sometimes found in rivers and streams or in coastal lagoons.

The most specialised predator of these rocky coasts, however, is the bass (*Dicentrarchus labrax*). Extremely fast and agile, with an enormous appetite, this species is the major enemy of shoals of small fishes and of crustaceans that scuttle in and out of the water.

Seashores of sand and mud

In coastal areas where sand and mud are predominant, the gentle mechanical action of the tides and the soft, ever-shifting nature of the substratum create conditions that are completely different from those prevailing on and around rocky shores. Although both forms of coastline are subjected to similar kinds of outside influence, survival on sandy and muddy coasts depends on

The sea horse is a small marine predator which captures victims, including fry and small crustaceans, by sucking them into its tube-like mouth. Larger prey are broken up by the force of the suction.

Facing page: Most predatory fishes of the coastal belt live in the open sea but come inshore to feed. One such species is the blue-spotted argus of tropical waters.

The torpedo or electric ray captures prey by delivering electric shocks from special organs situated in its wings (pectoral muscles).

successful adaptation to an environment where the yielding structure of the substratum prevents animals forming crust-like colonies or anchoring themselves firmly by a foot or similar extension of the body. Yet currents and tides are still strong enough to pose a continual threat to the organisms inhabiting these waters. In order to prevent themselves being swept away, their only means of self-protection is to burrow into the soft sand or mud. So these are conditions which favour a wide range of crawling animals.

Because there are so many more animals in these areas which rely on camouflage rather than on bright warning colours, zonation is less immediately obvious here, and at first glance it may be difficult to determine the frontiers of distinct regions. Nevertheless, such zones are actually far more extensive in these surroundings of mud and sand than are the equivalent zones found along rocky shores, for as a result of the more gradual incline of the substratum, the distances between high- and low-water marks are that much greater.

It is therefore possible to distinguish three zones, each of which is occupied by various species of crustaceans. The highest of these zones is characterised by the presence of amphipods (water fleas) which play the same ecological role as do periwinkles on rocky coasts. The predominant forms to be found in

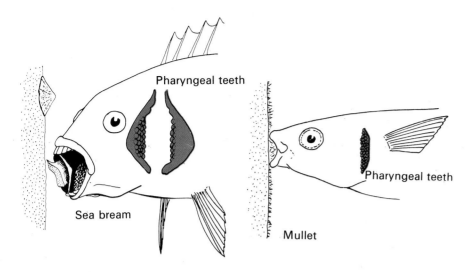

Sea bream and mullet have similarly modified mouth parts for satisfying their special food requirements. The former have sharp incisors for prising loose limpets clinging to rock faces; the latter have horny lips for dislodging algae attached to stones. Both species possess pharyngeal teeth for grinding up their food.

the intermediate zone are isopods; and the lowest zone is the habitat of other species of crustaceans, the amphipods.

The principal feature of these littoral formations is the lack of homogeneity in the substratum, the components of which vary in size from large pebbles and stones to tiny granules of sand or mud, this being dependent upon the action of the waves. Constituents of such particles include fragments of broken rock, notably quartz, scraps of shells (often partially eroded) and organic substances such as the bodily remains of animals and plants, and various forms of excreta. All these materials, in conjunction, stimulate the growth of a multitude of living creatures belonging to different biological groups—bacteria, diatoms, protozoans, rotifers, nematodes, turbellarians, hydrozoans and molluscs—together with larvae of other groups. Being subject to similar forms of adaptation, they tend to present a fairly

Although the sandhopper sometimes ventures into stony or seaweed-covered zones, other amphipods of the seacoast seldom leave the sandy beaches.

The starfish is one of the most formidable predators among marine invertebrates. Here a mollusc is trying vainly to escape from the starfish *Marthasterias* (*right*).

The colourful tube worms (*below*) extend the front part of their body which is furnished with a tuft of bristles. This is used for respiration and for stirring up currents which attract food particles.

Facing page: Most of the crustaceans of the coastal belt crawl along the bottom. The mantis shrimp has large claws for capturing in the manner of the insect, praying mantis, the small marine animals on which it feeds.

uniform appearance and they are generally about one-tenth of a millimetre long.

Of all the protective manoeuvres available to animals living in sand and mud, burrowing is that which is most frequently adopted. Many species of marine worms build themselves tube-like burrows in the substratum where they can take refuge at low tide. Commonest of Atlantic forms is the lugworm (*Arenicola marina*), widely used by fishermen as bait, whose tubular galleries are U-shaped and through which particle-laden currents circulate. This system allows the lugworm to feed (in a similar manner to the earthworm), for the mud and sand pass through the digestive tract, which retains only food particles. All such invertebrates are invisible at low tide, but when sea water flows in to cover their burrowing sites some of them (*Spirographis, Protula, Sabella, Serpula*, etc) extend the front part of their body, spreading out brightly coloured tufts of bristles. These have a double function, playing a part in respiration and stirring up miniature whirlpools which attract food substances. Among creatures also burrowing in this manner are acorn worms (*Balanoglossus* and *Glossobalanus*). Although they have a worm-like body they are not related to true worms. The presence of a special structure at one time believed to be the counterpart of the notochord (rudimentary spinal column) of the amphioxus has given them a position in scientific classification fairly close to that of the vertebrates.

Bivalve molluscs are perfectly at home in mud or sand. The

Hermit crabs often live in symbiosis with sea anemones. The association assures the crabs of security (they are immune to the poisonous stings) and the anemones of easy access to food.

two valves of the shell form a pressure-resistant chamber and the mobility of the foot or muscular extension of the body enables them to bury themselves quickly and efficiently. Among such bivalves are the common cockle (*Cardium edule*), the crosscut carpet shell (*Venerupis decussata*), the warty venus (*Venus verrucosa*) and striped venus (*Venus gallina*), the donax (*Donax trunculus*), the grooved razor shell (*Solen marginatus*), the common razor shell (*Ensis ensis*) and many other species. As a rule those bivalves burrowing near the surface tend to possess a strong shell that will withstand the strong lateral pressures exerted by the movements of the water-drenched sand. But

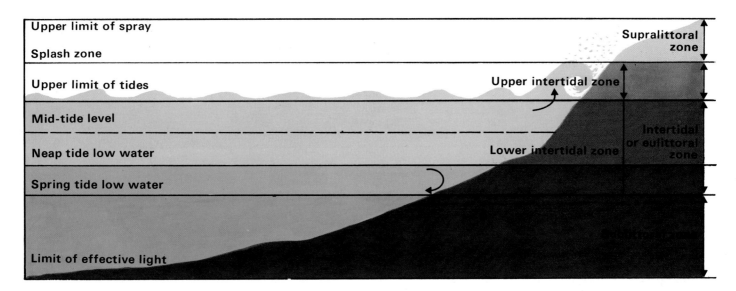

On the seashore, forming the frontier between two worlds, three major zones can be distinguished—the supralittoral, sprinkled by spray, the intertidal or eulittoral, where the sea level is continuously fluctuating, and the sublittoral, permanently submerged.

those digging themselves in more deeply are not subjected to such powerful pressures and are able to survive with a much thinner, more delicate, shell.

Certain echinoderms, such as sea urchins and even some large crustaceans (such as the crabs of the genus *Ocypode*) also exhibit similar types of burrowing habits.

Apart from amphipods and isopods, crustaceans are the most common forms of crawling animals. They include shore crabs (*Carcinus*), spider crabs (*Maia*), swimming crabs (*Portunus*), crabs of the genus *Pachygrapsus,* mantis shrimps (*Squilla*) and many others.

The ghost crabs of the genus *Ocypode* are to be found in enormous numbers on sandy tropical beaches. Some of them are very large, spending most of the day concealed inside deep vertical burrows, leaving so narrow a space between the walls that it would seem almost impossible for them to squeeze their body through. At nightfall they leave their refuges and venture out for food, principally in the form of vegetable debris that has been washed up on the shore. At dawn and dusk, therefore, a tropical beach extending for dozens of miles (as on the low-lying shores of Kenya and Tanzania, fringing the Indian Ocean) offers the astonishing spectacle of hundreds of thousands of these crabs scuttling to and fro on their long legs, stalk-like eyes providing periscopic vision in every direction. They snap up all available food and perform strange dancing movements (doubtless a form of intraspecific communication), waving their claws back and forth. The rhythm of the tides exerts a strong influence on their lives, especially when reproducing, for they lay their eggs in the sea. The fiddler crabs of the genus *Uca* display similar habits. These and other crustaceans are hunted by sea birds (gulls, terns, frigate birds, etc) and by various land mammals.

The ecosystem of the seashore, frontier of two worlds, is under the simultaneous influence of land, sea and air. The prosperity of such a specialised community depends on the presence of neighbouring zoological groups, both for the conveyance of organic substances (which are appropriately transformed) and for the population control exerted by the predators feeding on many seashore species.

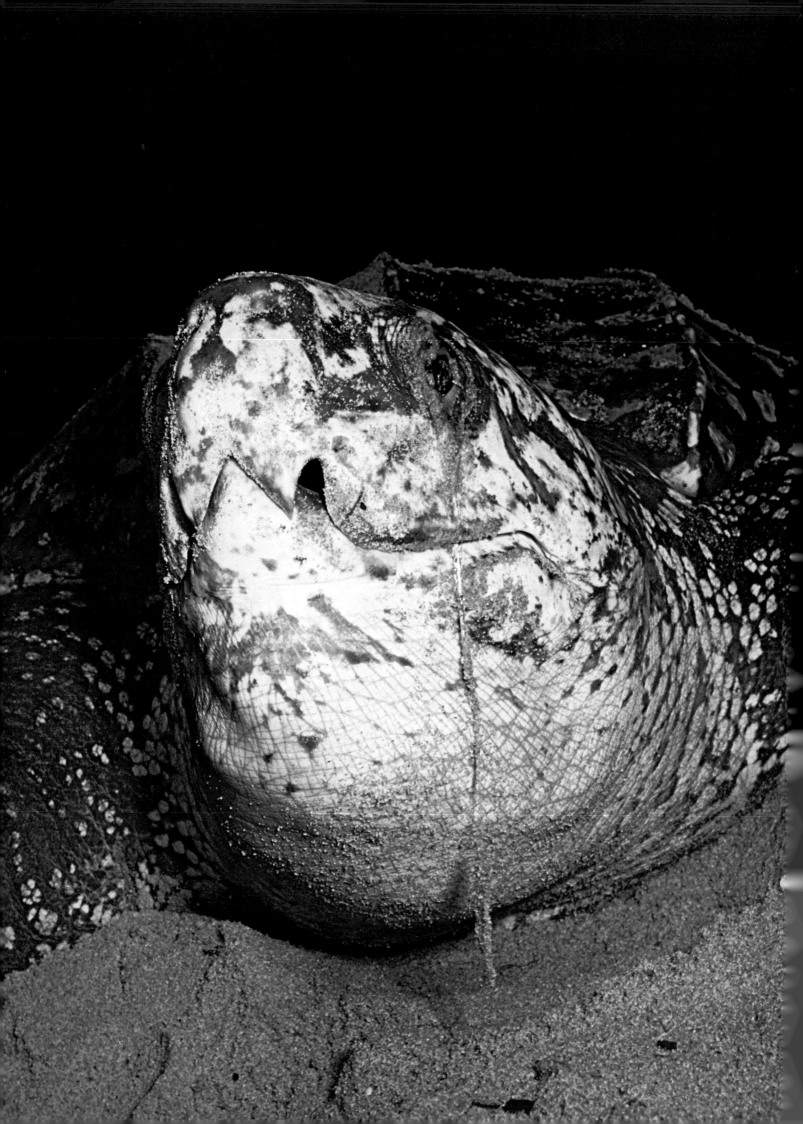

CHAPTER 3

Sea turtles and sea snakes

Reptiles, like all the large groups of terrestrial vertebrates, underwent complex adaptations in the course of their initial conquest of dry land but some of them, in a similar manner to certain mammals and birds, eventually made their way back to the sea. This return to ancestral waters, to the ocean basins where life itself originated, had something inevitable about it, as if the animals themselves were incapable of resisting the mysterious call of the deep.

The cetaceans, marine turtles, penguins and sea birds which spend almost their entire life in a watery environment, are the most perfect examples of an evolutionary process which entailed a complete restructuring of the organism. All were animals which already breathed with the aid of lungs, whose muscles were attached to a heavy skeletal frame and which had lost the hydrodynamic forms associated with marine locomotion.

At the peak of their development, the reptiles included sea-going species as perfect as the ichthyosaurs, extraordinarily like dolphins and doubtless with comparable swimming capacities. There were also veritable sea dragons such as the elasmosaurs, which measured 50 feet long, with snake-like neck and formidable teeth that must have made them the terrors of the ocean.

Along with the other dinosaurs these giants of the seas vanished completely, for reasons that are not understood, leaving only a few bones as evidence that they ever existed. Just as inexplicably, however, two groups have survived to this day, namely the sea turtles and the sea snakes. The former particularly seem to have retained something of the power of their ancestors, with many of the primitive characteristics and much the same overall dimensions that were the hallmarks of the marine reptiles of the Secondary era.

Facing page: When the time is ripe, the female leathery turtle leaves the sea and clambers up a sandy beach to lay hundreds of eggs. The secretions flowing from her eyes are not, as has been popularly believed, tears but the products of ocular glands which help to counteract the dryness of the air.

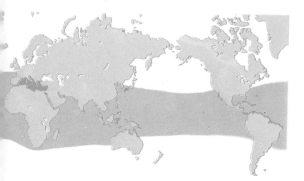

Geographical distribution of the green turtle.

Witnesses of vanished glory

Few animals are as impressive as the leathery turtle or leatherback (*Dermochelys coriacea*). Weighing more than half a ton and measuring as much as 12 feet with flippers fully extended, this giant is, as might be expected, a powerful swimmer, only coming to the surface at intervals to breathe. When the time comes for her to lay her eggs, the female clambers slowly and laboriously up the steep slope of a tropical beach. Having deposited her eggs in a suitably dry spot, well beyond the reach of the waves, she ambles back into the sea, vanishing into the depths. Yet year after year the females will be back on the identical sandy shores, ready to repeat the procedure. It is a remarkable accomplishment, considering the profound transformations that the turtle's ancestors underwent in adapting once more to life in the sea, after a lengthy period on dry land. The sea otters, the seals, the huge cetaceans and the great auks (now extinct) embarked on similar adventures. To have survived at all was a miraculous achievement.

As far as the leatherbacks and other marine turtles were concerned, adaptation to an ocean environment proved no guarantee of a long and carefree future. Their appearances on tropical shores, far from evoking admiration and curiosity (except among naturalists) aroused only the baser human instincts of greed and cruelty. For centuries nests have been pillaged, eggs stolen and defenceless females slaughtered. Breeding sites have been turned into graveyards.

The five extant species of marine turtles live in tropical and subtropical seas. Palaeontologists are satisfied that the carapace of modern species is virtually the same as that of primitive ancestral forms, although to some extent modified. How such a shell developed and when the earliest turtles appeared, they do not know. The oldest fossil remains have been discovered in sedimentary rocks of the Upper Cretaceous period, the time when the largest known tortoise, *Archelon*, flourished. But it is generally believed that even more primitive forms existed in the Triassic.

Both freshwater and marine turtles are streamlined in shape and their carapace is much flatter, more lightweight and less tightly sealed than that of land tortoises. But an even more significant difference is that the feet of the turtles which spend the greater part of their time in water have gradually been transformed. The process is most striking in the case of the sea turtles, whose limbs have been changed into flippers and whose claws have atrophied.

The surviving marine turtles are divided into two familes. The Cheloniidae comprise four hard-shelled species—the loggerhead (*Caretta caretta*), the hawksbill (*Eretmochelys imbricata*), the green turtle (*Chelonia mydas*) and the Kemp's Ridley turtle (*Lepidochelys kempi*). According to some authors there are two species of *Lepidochelys*, but others recognise only two subspecies. The Dermochelyidae have only one living reprsentative, the leatherback, which is the largest of all turtles.

Strange as it may seem, considering their size and relative

GREEN TURTLE
(*Chelonia mydas*)

Class: Reptilia
Order: Testudines
Family: Cheloniidae
Length of carapace: male $38\frac{1}{2}$ inches (97·6 cm)
female 33 inches (84·2 cm)
Weight: 528–726 lb (240–330 kg)
Diet: basically herbivorous (underwater plants) but occasionally fishes, molluscs and crustaceans
Number of eggs: 220–500
Diameter of egg: about $1\frac{3}{4}$ inches (4·5–4·6 cm)
Incubation: 40–72 days

Back greyish-brown with olive reflections. Upper part of head bright chestnut with edges of scales yellowish. Carapace elongated, narrowing towards the rear more markedly in male than female. Belly whitish, except in parts where body and legs meet, these being blackish, lightly flecked with white.

Loggerhead (*Caretta caretta*)

Green turtle (*Chelonia mydas*)

Hawksbill (*Eretmochelys imbricata*)

Leatherback (*Dermochelys coriacea*)

abundance, little is known of the biology and behaviour of these marine turtles. The lack of information stems largely from the fact that they spend almost their entire life in the water, often below the surface. Furthermore, zoologists have only recently begun to take a real interest in them.

Reproduction is the only phase of the sea turtle's life cycle which has been observed in detail, for this spectacular, even dramatic, activity takes place on land. When the time draws near for egg-laying, the females return to the tropical beaches. The sites are always the same, year after year. Most probably this breeding behaviour is a kind of genetic 'imprinting', the information concerning place and timing being transmitted from generation to generation within a species, even within a given population.

On certain shores the local people profit from these ancestral behaviour patterns to kill the females, not when they first set foot on land but after they have laid their eggs. Each female lays approximately 400 eggs, but at intervals and in different places. Dispersing the eggs in this manner is undoubtedly a

safety measure designed to avert the possible catastrophe of losing a large part of, or even all, the eggs to land predators. While the females are laying their eggs, the males assemble on the same beaches, waiting until their partners return to the water before copulating with them. This curious habit, rare among other vertebrates, arises partly from the fact that the spermatozoa of turtles are extraordinarily potent, remaining alive in the genital tracts of the females for many months and sometimes being active for several years after copulation occurs. It is therefore possible that a newly laid egg will be fertilised by a spermatozoon received on a previous mating. But even stronger reasons for this unusual form of mating behaviour are the needs to conserve energy and to guarantee security. Consequently males and females couple once only and in a particular spot where the sexual act and egg-laying can take place virtually simultaneously. Marine animals which reproduce on land, such as the pinnipeds (seals and walruses) have, by a process of convergent evolution, similarly compressed mating and giving birth into the shortest possible period, in their case thanks to an exceptionally long gestation.

So far no real light has been shed on other aspects of sea turtles' behaviour—especially the means by which they find their way through the vast uncharted ocean from the regions where they normally live, back to the beaches where they lay their eggs. How do they select the routes and what methods do they use for navigating so unerringly to their chosen destinations?

Karl von Frisch's classic studies of the behaviour and communication of bees have revealed that honeybees are guided by the position of the sun in relation to an invisible line running from the hive to the place where they gather food. G. Kramer and others have since shown that migrating birds also navigate by means of the sun and stars. So it has been suggested that marine turtles find their way through the oceans in a similar way, although so far there has been no absolute proof of this. The theory has also been put forward that the reptiles make use of their sense of smell, in the same manner as salmon, to identify the waters through which they travel, but this again is no more than hypothesis. As for the claim that they are guided by a form of echo-location (analogous to that of cetaceans), this must remain highly doubtful, at least until opportunities occur to determine whether the turtles possess a well developed sense of hearing.

Although marine turtles normally leave the ocean only to lay their eggs on the beaches, green turtles and occasionally loggerheads may from time to time bask in the sun on islets or on rocky promontories.

The relationship between man and the various species of sea turtles is a sad example of the disastrous consequences of greed and ignorance which in so many parallel cases has led to the massacre, sometimes extermination, of a zoological group. For centuries the inhabitants of regions where turtles gather for breeding have eaten the flesh and eggs of the reptiles, but have taken only as much as they need so as not to jeopardise future

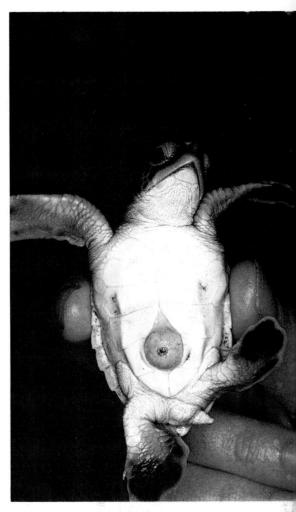

The vitelline sac is clearly visible on the belly of this one-day-old green turtle.

Facing page: Having laid her eggs in various sand hollows, the female green turtle returns to the sea and pays no further attention to them. After an incubation period ranging from forty to seventy-two days the babies hatch, almost at the same time, and make their way as quickly as possible down to the water.

From birth until they are ready to reproduce green turtles hardly ever leave their watery environment. Although little is known of their habits it has been established that the adults, which live in the open ocean, are herbivorous, whereas the young at first feed on invertebrates caught near the shore.

supplies. No such commonsense has motivated the activities of the industrial concerns which have more recently discovered the huge profits to be made in canning turtle soup (obtained from the cartilaginous substance filling the interstices of the plastron). It did not take long for this trade to destroy the age-old equilibrium of the turtle population. The demand and, as a result, the need to fulfil it, has become so great that entire communities of turtles have been massacred in the breeding season, bringing about an alarming decline in numbers. Today, unless this insane traffic is halted or controlled, there is a real danger of sea turtles disappearing for ever. It is all too reminiscent of the sad plight of the large whalebone whales and of the blue whale more especially.

Hatching of the eggs only takes a brief time and occurs almost simultaneously in every colony. Consequently hordes of newly hatched turtles may be seen heading blindly for the sea. This is the most critical stage in their life. Completely defenceless, the tiny reptiles inevitably attract a number of land predators, including sea birds, and sometimes only a very small proportion reach their destination.

The green turtle

The green turtle is probably the best known of the marine species, both because of its abundance and from the fact that it is the principal source of turtle soup. The majority of green turtles live in comparatively shallow tropical waters, generally in the open sea, where the rocky bottom is well covered with algae which also provide places of refuge.

Essentially herbivorous, the green turtle feeds chiefly on marine plants of the genera *Zostera* and *Thalassia,* but occasionally supplements its diet with molluscs and crustaceans. Zoologists sometimes distinguish two subspecies, one an inhabitant of the Pacific, the other of the Atlantic, which are additionally differentiated by colour.

The activities of green turtles have been studied more successfully than other species by reason of the fact that they swim close to the surface. But they too have thus far retained their secrets of direction-finding. How they make their way back to their breeding zones, situated as they are many thousands of miles from the oceanic islands which are their homes, remains a complete mystery, although perhaps it is by celestial navigation.

The breeding season varies considerably according to where the different communities live. Those that lay their eggs away from tropical latitudes tend to concentrate all their sexual activity into the three hottest summer months. On the other hand, the green turtles of the tropics lay their eggs all the year round. Incidentally they are among the rare reptiles which do not reproduce regularly each year. In Borneo, for example, the female lays eggs once every third year, in Central America in alternate years.

As is the case with other sea turtles, the journey from the water's edge to the nesting site is long and arduous. The enormous animal drags herself heavily up the beach, leaving a deep furrow behind her. From time to time she comes to a halt, the sounds of her noisy breathing being rather like a series of heavy sighs. This breathlessness is understandable, for a really tremendous effort is required for these animals, out of their natural element, to fill their lungs with air. Furthermore, because the air is so dry, the ocular glands of the turtles exude a thick, translucent liquid, popularly, but mistakenly, supposed to be tears of grief!

Having selected a convenient nesting site, each female excavates a hole with her hind flippers, into which she deposits the eggs (looking like ping-pong balls). She then covers them with sand and carefully flattens the surface. Now unburdened, and taking advantage of the descending slope of the beach, she makes her way rapidly back to the ocean. But she will repeat the process more than once, in fact between two and seven times, at approximately two-week intervals, until she has laid about 400 eggs. The incubation period ranges from 40 to 72 days, according to the latitude. How many will hatch is largely a matter of luck, for the various nests may be ransacked by rodents and dogs as well as by man.

Copulation takes place after all the eggs have been laid,

Difficult to take at sea, marine turtles are usually killed or caught in the vicinity of their breeding sites. Hunted both for their shells and for the nutritious substance that forms the base of turtle soup, almost all species are in danger of becoming extinct.

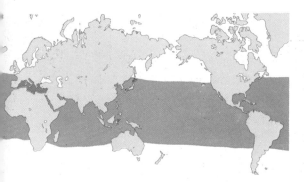

Geographical distribution of hawksbill and loggerhead.

HAWKSBILL
(*Eretmochelys imbricata*)

Class: Reptilia
Order: Testudines
Family: Cheloniidae
Diet: omnivorous (crustaceans, molluscs, algae)
Number of eggs: about 500
Diameter of egg: about 1½ inches (3·5–4·1 cm)

General colour amber, with russet, dark chestnut or yellowish marks. Plastron yellow, flecked with black, especially towards front. Head narrow, jaw hooked at tip.

LOGGERHEAD
(*Caretta caretta*)

Class: Reptilia
Order: Testudines
Family: Cheloniidae
Total length: up to 51 inches (130 cm)
Weight: 726–946 lb (330–430 kg)
Diet: essentially carnivorous; some vegetation
Number of eggs: 120–150
Diameter of egg: 1½–1¾ inches (4–4·3 cm)
Incubation: 30–75 days

Large turtle with broad neck. Back russet-chestnut. Plastron yellowish.

Facing page: The beautiful carapace of the hawksbill (*above*) is much sought after in the East for manufacturing ornaments. Its habitat often overlaps that of the loggerhead (*below*) but it is easily distinguished from the latter by the shape of the shell.

usually in the open sea, far from the shore. It is quite common to see two males accompanying each female. When the incubation period is over the babies break the shells of the eggs with the aid of a horny egg-tooth on the tip of the upper jaw, and burrow out of the sand. By some miracle of timing, they all hatch more or less simultaneously, which means they emerge from the sand in large groups. Their forelimbs are already well developed and they lose no time in scuttling as fast as they can towards the sea. Those that manage to evade predators and reach the ocean do not swim out very far at this early stage of life, but feed principally on invertebrates caught near the coast. They are seldom seen in company with adults which normally feed far from the shore. The only place where they have been observed in large numbers is on the west coast of Florida, where they gather in the spring. Here they remain until the autumn. But before striking out for more distant waters they change their diet, concentrating mainly on aquatic plants. Then, either in late autumn or early winter, they begin their migration, disappearing without trace in the open seas and not returning to land until they are adults.

The hawksbill

The hawksbill is easily identified by its attractive shell, made up of multicoloured, translucent plates. This carapace forms the basis of the lucrative tortoiseshell industry of the Orient, notably in Ceylon, Indonesia and Japan.

This species inhabits tropical seas but unlike the green turtle is generally to be found in bays, river estuaries and any areas of shallow water with scattered plant cover. Although its distribution may sometimes coincide with that of the green turtle and the loggerhead, it is not usually present in such numbers. Of medium size (the shell of an adult does not measure more than 3 feet in length), this turtle is omnivorous, though there is a preference for substances of animal origin. The diet consists fundamentally of crustaceans, molluscs and algae.

Experts sometimes distinguish two subspecies, again with one race in the Pacific and the other in the Atlantic. The former breeds from November until February, the latter from April to August. Eggs are laid in two or three stages, at intervals of less than three weeks, until about 500 have been deposited by each female. At birth the individual shields of the carapace overlap one another but as the baby turtle grows these gradually lock into the normal position.

The loggerhead

Considerably larger than the green turtle is the loggerhead, which differs from other marine turtles by reason of an elongated carapace with characteristically marked external plates. This species frequents shallow bays, sometimes appearing in river estuaries, but is seldom observed on the high seas.

Basically carnivorous, the loggerhead feeds in the main on fishes, molluscs, crustaceans and sponges. But analyses of

The greatest threat to the lives of newly hatched turtles comes immediately after they hatch and scuttle towards the ocean.

stomach contents have revealed the presence of a certain amount of plant matter, notably of the genera *Zostera* and *Thalassia*.

The range of distribution of the loggerhead extends from the northern to the southern limits of the tropics and here too there are two distinct subspecies in the Pacific and Atlantic Oceans. The Pacific race breeds along the entire length of the Californian coastline, the Atlantic subspecies off the shores of Florida and Georgia. The females lay their eggs (totalling 120–150) in May, June and July, although the first batches, according to reports of local fishermen, usually appear at the time of the first spring tide in June. As a protection against the tides, the eggs are laid beyond the high-tide mark. The incubation period is between 30 and 75 days.

The Kemp's Ridley turtle

Until comparatively recently herpetologists regarded the populations of the Kemp's Ridley turtles inhabiting the Atlantic and the Pacific/Indian Oceans as two distinct species, the Kemp's Ridley and the olive Ridley. Today, however, they are classified as subspecies of *Lepidochelys kempi*. What distinguishes them, above all, from green turtles, are the five pairs of lateral shields on the carapace.

Much prized for its flesh and eggs, the Kemp's Ridley turtle, once abundant in the Pacific, the Indian Ocean and the Atlantic, is today a threatened species.

The Kemp's Ridley turtles are the smallest of the sea turtles, for the carapace of the adults does not exceed 28 inches in length. Among Florida fishermen it is popularly believed that they are crosses between green turtles and hawksbills, but this seems unlikely in view of the fact that they are distinguished from these two species by their marginal plates.

These turtles generally inhabit shallow waters. Their diet is basically vegetarian, supplemented occasionally by molluscs and crustaceans.

The populations found in the Caribbean reproduce between December and February, those of the eastern Pacific from August until November, those from the waters around Ceylon from September to January, and those living along the coasts of Burma either in March or April.

The turtles are so cautious and shy that it has proved difficult to observe their breeding rituals. One female, however, was sighted leaving the water at nightfall and making her slow journey, guided by instinct, towards the spot chosen as a nest. She progressed by moving first one front flipper, then the other, stopping every three steps or so to rest. Having reached her destination she scooped a hole, a little more than a foot deep, and proceeded to lay her eggs in batches of two or three, at

The leathery turtle, seemingly crushed by her own weight, digs a hollow in the sand to lay her eggs. These and other species, so agile in water but clumsy on land, are cruelly exposed at such times. Because of uncontrolled hunting they are becoming increasingly rare and some are on the way to becoming extinct.

Geographical distribution of marine turtles.

Facing page: Recognisable by the longitudinal ridges along the carapace, a female leathery turtle crawls back, after laying her eggs, to the sea, leaving behind her marks similar to those made by a caterpillar tractor.

KEMP'S RIDLEY TURTLE
(*Lepidochelys kempii*)

Class: Reptilia
Order: Chelonia
Family: Cheloniidae
Length of carapace: 24–28 inches (60–80 cm)
Diet: vegetation; also molluscs and fishes
Number of eggs: 300–400
Diameter of egg: 1½–1¾ inches (3.8–4.3 cm)
Incubation: 50–60 days

Colour of shell varies according to individual, ranging from dark grey to olive-green. The plastron is usually whitish-yellow. The two subspecies are *Lepidochelys kempii* (Atlantic) and *L. kempii olivacea* (Pacific).

LEATHERY TURTLE
(*Dermochelys coriacea*)

Class: Reptilia
Order: Chelonia
Family: Dermochelyidae
Total length: 63–91 inches (160–230 cm)
Weight: 990–1320 lb (450–600 kg)
Diet: omnivorous
Number of eggs: about 300
Diameter of egg: about 2¼ inches (5.5–5.8 cm)
Incubation: 55–65 days

Dark chestnut or black carapace, flecked with white, narrowing towards rear and divided by seven longitudinal ridges. Plastron whitish with a few irregular black marks.

---

four- to six-second intervals. When finished she carefully covered the nest with sand and returned to the sea. The entire operation took about an hour.

The leathery turtle: giant of the tribe

The leathery turtle or leatherback is the only representative of the family Dermochelyidae. Although the distribution range of the species extends to the warm waters of many oceans, individuals normally tend to confine themselves to a particular region. Just the same, leatherbacks sometimes make long journeys through the high seas to appear off the shores of the British Isles, Spain, South Africa and Argentina.

Of all turtles, including freshwater species, they are the best suited to aquatic life. Largest of the whole order, the adults may weigh more than 1,200 lb and measure up to 7½ feet. In general appearance they look much like other turtles but their forefeet are transformed into enormously powerful flippers. The carapace too has undergone modification, having become elongated and smooth, possibly to facilitate passage through the water. It is comprised of small, juxtaposed bony shields of various shapes, which are lying under the skin, and it is marked by seven longitudinal ridges. This type of armour-plating is thus different in structure from the shells of other species, giving rise to a number of theories as to their origin and relationship to the Cheloniidae. Some authors believe that they may be derived from a group of giant land turtles that later adapted to life in the ocean and which resemble other marine species only as a consequence of convergent evolution. Other experts suggest that leathery turtles are considerably older than related living species, that they evolved over a much longer period and that for this reason they acquired the most effective forms of adaptation to marine conditions.

Very little is known of the habits of these turtles. Nevertheless, those zoologists who have been able to study them in some detail are agreed on three points. They are extremely strong and, as a result, are likely to represent quite a danger to anyone attempting to capture them, for they are capable of defending themselves energetically with their front flippers and with their powerful horny snout. Secondly, they give out characteristic sounds; and lastly, they swim very rapidly, having been timed at around 10 seconds over a distance of 100 yards.

As far as breeding behaviour is concerned, leatherbacks do not differ notably from other species. The season varies in accordance with latitude. On the coasts of Florida December and January are the peak months. Off Honduras and Nicaragua reproduction takes place from May to August; and in the Indian Ocean in May and June. Eggs are laid in two or three batches, with each female eventually depositing about 300. Each egg is round, white and just over 2 inches in diameter.

The female digs a hollow in the sand with her hind flippers, covering the eggs in the usual manner for concealment and warmth. Although she displays extreme caution when emerging from the sea and hauling herself up the beach to the nesting site, she

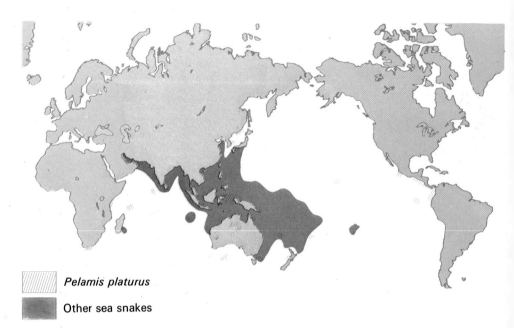

Pelamis platurus

Other sea snakes

Geographical distribution of sea snakes.

appears to lose all sense of danger when actually laying the eggs, exhibiting no reaction to noise, light or even the presence of humans.

Once having unburdened herself, the female leatherback follows her former tracks back to the sea. Even now, however, she may be so exhausted by her efforts that she will be compelled to halt at intervals to take breath.

The babies hatch after an incubation of 55–65 days. They are born with carapace already developed and with scales all over the body.

Little is known about the feeding habits of these huge turtles. But analysis of the stomach contents of several individuals has revealed a virtually omnivorous diet, including fishes, crustaceans, algae, eggs and young octopuses.

The colourful sea snakes

During the Triassic, some 200 million years ago, coinciding with the appearance of the first dinosaurs, another event took place on earth which marked an important phase in the evolution of reptiles – the emergence of the snakes.

Modern herpetologists incline towards the theory that the closest relations of the snakes are the lizards, even though it has not been satisfactorily explained how the former came to lose their limbs nor how the two groups acquired a different eye structure. It leaves open the question of whether snakes evolved from limbless saurians or whether they were descended from reptiles with well developed limbs (similar to modern monitors), but now extinct.

Whatever the origin of snakes, it is obvious that the lack of, or loss of, limbs, together with the elongation of the body, directed the members of this group towards a completely new method of locomotion.

In the course of time, and doubtless for reasons of mobility, a number of reptiles took to life in the sea, where they found it much easier to move around, principally by swimming. But one of the chief reasons for this return to a marine environment was the abundance of food to be found in the ocean. Among individuals that made their way back to ancestral haunts were the dolphin-like ichthyosaurs, the plesiosaurs (resembling long-necked turtles), the mosasaurs and the teleosaurs or sea crocodiles. All these giant reptiles have long since vanished, for unexplained reasons. The only living testaments to their former dominion are the marine turtles and sea snakes.

There are approximately 50 species of sea snakes, distributed widely through the Indian and Pacific Oceans. Because of their special anatomical structure they are classified in a separate family, Hydrophidae. They differ from terrestrial snakes in several aspects. The head is small and the neck and front part of the body are slender in comparison with the thicker abdomen. The tail too is modified, being flattened from side to side and thus more suitable for swimming. Another characteristic feature of the sea snakes is that the nasal pits are directed upwards, enabling the reptiles to breathe without raising the head completely out of the water. Most of them measure 3–4 feet, but some are larger.

Although the snakes comprising this family possess these and other common features which make them a relatively homogeneous group, there are certain differences which divide them into two distinct subfamilies. Among the Laticaudinae the scales of the back are small and neatly overlapping, whereas those on the belly are large, more resembling the scales of terrestrial snakes. They include some of the most primitive of the sea snakes, notably those of the genera *Laticauda, Emydocephalus* and

Sea snakes, abundant in the tropical waters of the Pacific and Indian Oceans, are all venomous, feeding mainly on fishes. The females give birth to live young.

Aipysurus. One species, *Laticauda semifasciata*, measures up to 6 feet long and 3 inches in diameter. Found in large numbers in the waters off the Philippines, it is hunted both for its beautiful skin and for its appetising flesh.

The representatives of the second subfamily, Hydrophinae, are better adapted to conditions of life in the open sea, and differ from the Laticaudinae in that the scales of the abdomen are so reduced in size as to be almost identical to those of the back. As a result, these snakes can slither more easily through the water. Snakes of this group tend to be large—those of the genus *Hydrophis*, for example, measuring close to 8 feet in length. Furthermore, the diameter of the belly is noticeably greater than that of the neck. In the case of average-sized species it is roughly double, but it is about four or five times as thick in *Microcephalophis gracilis*.

One important characteristic of most sea snakes is that the females are viviparous. This means that they can remain in the water throughout the entire reproductive cycle.

Sea snakes are generally found in comparatively shallow water where fishes are abundant, and they show a marked preference for eels.

Sea snakes seldom venture out of the water, but members of the subfamily Laticaudinae in particular may sometimes appear on shore, searching for the eggs of sea birds.

The yellow-bellied sea snake (*Pelamis platurus*), brown or black on the back, with a yellow belly, has the widest distribution of all. This pelagic species ranges across the Pacific and has been sighted off the west coasts of America. Sometimes large groups are seen hunting on the surface. Their favourite technique is to remain completely motionless, simulating a floating object. Clouds of small fishes which have the habit of exploring logs, bits of wood and other debris then swarm round, unaware of the danger they are courting.

The toxicity of venom varies according to species and, in some measure, in relation to the habitats they occupy. In rough seas, reefs and submarine prairies it is essential that the poison should be active and take immediate effect. If not, the victim, carried away by strong currents, or taking refuge among the tangles of seaweeds, may not die until it is well beyond the reach of its predator. In calmer, shallower waters, however, there is no need for the venom to be so powerful, merely strong enough to prevent the victim getting clear away before drawing its last breath.

Many sea snakes venture into river estuaries but there is only one known species which has adapted to life in fresh water. This is *Hydrophis semperi*, found in Lake Taal on the island of Luzon in the Philippines. The lake covers the crater of an ancient volcano and has an outlet to the sea a few miles away through which migration was possible.

As a rule, sea snakes do not travel any great distance in the ocean, the exception being the afore-mentioned yellow-bellied sea snake. This is probably the principal reason why the sea snakes of Asiatic waters have remained quite separate from those living off the Australian shores.

Facing page: The flat-tailed sea snake (*Laticauda colubrina*) occasionally ventures onto dry land to supplement its diet with the eggs of sea birds. The female, almost twice as large as the male, measures up to 56 inches.

YELLOW-BELLIED SEA SNAKE
(*Pelamis platurus*)

Class: Reptilia
Order: Squamata
Family: Hydrophidae
Diet: small fishes

Dark brown to black above and bright yellow below, the two colours being separated by a well defined line. Tail yellow with large black marks. Female viviparous. The only truly pelagic species of sea snake, ranging across Pacific to west coasts of America.

CHAPTER 4

The birds of the coastal fringes

Ecologists have been able to demonstrate quite convincingly that transitional zones between two different types of vegetational zone are rich both in species and number of individuals. Thus where forest and savannah merge, for example, multitudes of animals tend to gather, making use of the natural resources of both environments to feed, shelter and reproduce.

Of all the frontiers dividing two habitats, none can be more definite and clear-cut than the coastline separating land from sea. From the broad belt extending from rocky shores or sandy beaches (already subjected to the influence of the ocean) out into the middle depths of the coastal waters, life is extremely diversified, mainly because these are areas that are both covered by water and directly reached by sunlight. Enormous numbers of invertebrates and vertebrates live alongside one another in this coastal fringe, and the interactions between producers, primary consumers, predators and scavengers go to make up a complex and fascinating patchwork of life.

The birds of the seacoast, despite the demanding physical and physiological adaptations necessary for flight, show remarkable versatility and flexibility, and this evolutionary faculty has resulted in the creation of a large number of species. For the ordinary bird lover, but to an even greater measure for the ornithologist, these birds are of tremendous interest. For the sake of convenience, their world can be broken down into three distinct zones, each of which is exploited by particular groups and may, therefore, be dealt with in turn. The groups concerned are, firstly, the birds which obtain their food on the seacoast proper, that is on the beaches (whether or not covered by water) or on the cliffs; secondly, those species which procure food from the surface of the sea; and, thirdly,

Facing page: Guillemots (seen here), razorbills and other sea birds form large breeding colonies on rocks battered by the waves. Because of numbers and difficulty of access they are relatively secure from predators and can lay their eggs and rear their young without risk of disturbance.

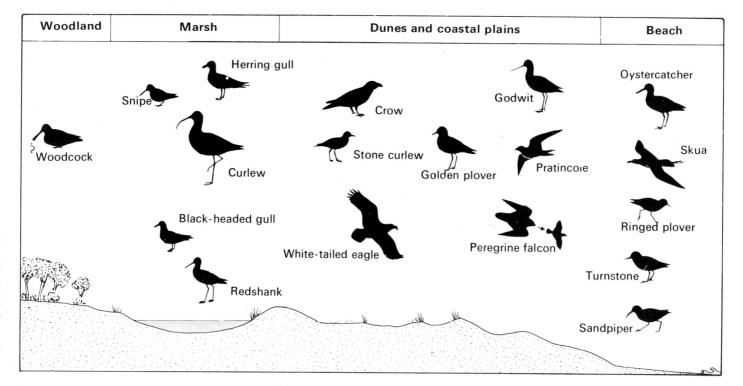

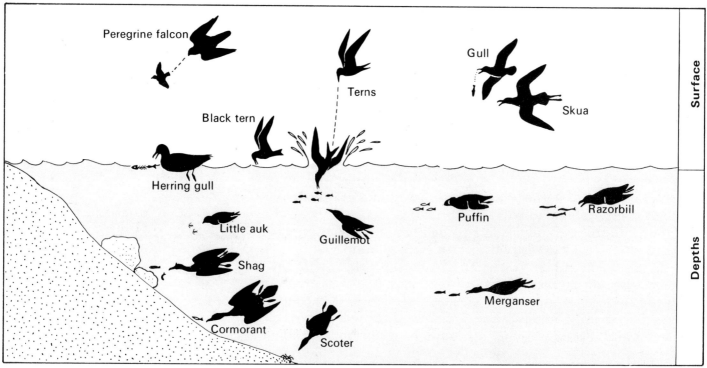

These diagrams show approximately some of the principal birds of the coastal fringe which are distributed on the shore. Their habitats range from woodland to beach, the various species hunting either in the air, on the surface or under water.

birds which capture their prey by diving into the water. In addition there are a number of non-specialised birds that hunt in all three areas simultaneously.

In the first group of birds, frequenting the shores, special mention must be made of waders that include sandpipers, redshanks, godwits, golden plovers and ringed plovers. These birds strut up and down, pecking at the sand and mud, rummaging in the bunches of seaweed thrown up by the tide, occasionally venturing down to the water's edge, without getting their plumage wet, to scoop up tiny marine animals. Other species, such as the oystercatchers and turnstones, probe for small invertebrates under the shingle and, thanks to exceptionally long legs, are able to wade into the shallow water.

The second group comprises those birds whose domain is the area of beach regularly bathed by the tides, and out into the sea for a distance of about 100 yards – the terns. All species of terns, wherever they are found, employ identical fishing techniques, diving down and catching with their beak any fish or invertebrate that happens to appear at the surface. The victim may struggle for a few moments before being swallowed by the tern in mid-flight.

Some distance beyond the line of the breakers, hunting in rather deeper water, are the sea birds of the third group, fusiform in shape, most of them swimming with paddle-like movements of their wings and diving below the waves for their fishy prey. They include guillemots, razorbills, puffins and cormorants.

Finally there are the gulls, the true opportunists of the seacoasts, equally at home on land and at sea, assiduously pecking into rock cavities, swooping gracefully over the sands or landing on the waves, ever on the lookout for suitable food. Since they have no exclusive speciality, whether walking, flying or swimming, the many species of gulls have the advantage of being able to exploit most of the food opportunities offered to the three preceding groups.

The reason that such a multitude of sea birds can live together in comparative harmony is precisely because the majority of them are specialists. Each group has well defined periods of activity, a selected range of prey, a chosen sphere of operations and characteristic hunting methods. Areas left vacant are automatically taken over by gulls and other species with an eclectic diet.

The abundance of food in this coastal fringe, coupled with the fact that it is concentrated into a relatively restricted space, is the principal reason for the varied nature of bird life on and around the seashore. Each species has been obliged to tackle the problems of life in a prosperous but circumscribed habitat in a different way. As in similar environments – the transitional zones between forest and savannah, between hill country and plain, between mountain and steppe – the diversification of species has been the key to survival.

The vast majority of birds inhabiting coastal regions belong to the order Charadriiformes. But although it might appear that the representatives of this very large order have monopolised the resources of this bountiful domain, they do in fact share their fortune with various birds of other orders, including ducks (Anseriformes), shearwaters (Procellariiformes), rails (Gruiformes) and cormorants, boobies, gannets and the like (Pelecaniformes), all of which frequent shores and coastal lagoons. Many of these birds, however, can only really be regarded as occasional passing visitors, taking temporary advantage of the food opportunities of this frontier zone.

Each of the principal groups of Charadriiformes has prospered by adapting in its own characteristic way to the changing conditions of seashore life. A detailed description of their diverse adjustments to the environment would naturally fill a separate book. All that can be attempted here, therefore, is to provide a general picture of their varied life styles by dealing in turn with the major species of the shore, the surface waters and the depths.

Terns owe their alternative name of sea swallow to their long, slender wings, forked tail and short feet. They are remarkable fliers, as proved by the long migration journeys of Arctic terns to the southern hemisphere and back.

Sea birds of sandy beaches, mudflats and rocks

The numerous species of birds that live and feed on and around the shore (not normally entering the sea) run a remarkable gamut of shapes and sizes. Among the main distinguishing features are the forms of beak and the structure of legs. On the basis of such anatomical characteristics alone it is possible to differentiate a series of smaller ecological groups.

In the first place there are those birds which, by and large, find their food in soft ground. Although they exhibit varying degrees of specialisation, their most prominent feature is a long bill, enabling them to probe deeply into sand or mud in order to extract invertebrates (generally worms). In this group, species such as the woodcock have chosen to live among trees, and some, such as the snipes, to conceal themselves in areas where rushes and sedges grow. Others, however, including godwits and redshanks, prefer to roam the sands and mudflats where the ground is either completely bare or at best covered with sparse vegetation.

Because they procure food in exposed places where there is little or no natural protection, such birds have acquired particular attributes to assist them in their daily activities. Thus many species have eyes situated in the upper part of the head, giving them the double benefit of being able to peck continuously at the ground, yet at the same time commanding a view of the surroundings, so as not to be surprised by potential predators. Furthermore, they appear to possess exceptional hearing, enabling them to pick up the slightest vibrations of their subterranean victims. To this end the ear openings are much farther forward than they are in most birds, in fact approximately midway between eye and beak.

At certain times of year frost may be a grave handicap to the way of life of these long-billed birds. They are particularly sensitive, therefore, to sudden drops of temperature, and this seems to give them advance warning so that they can lose no time in winging their way southward before winter arrives.

Other waders find their food in similar areas but on the surface of the sand or mud. The golden plovers and ringed plovers, for example, possess a bill that is somewhat shorter and more robust than that of the afore-mentioned species. Their legs, with rather short, sturdy toes, are better adapted for running; and in fact sprinting is the essence of the astute hunting technique of the ringed plovers. They race along behind the outgoing waves and scoop up any creatures left stranded on the beach, beating a quick retreat as the tide advances once more. The golden plovers, for their part, whose food consists in the main of worms and crustaceans, collected at low tide, are able to crack the shells of crabs, an aptitude carried to even greater lengths by the specialised crab plover (*Dromas ardeola*) of the coastal regions of Arabia.

Large molluscs which bury themselves in the sand or which attach themselves firmly to rock faces might appear to be fairly safe from such sea birds, but there are in fact a number of

Two examples of wading birds, the sanderling (*above*) and the black-tailed godwit (*below*). The former scurries over the sand, keeping pace with the retreating tide, probing for tiny marine organisms. The latter, thanks to its long legs, wanders into the shallow water and rummages at leisure in the mud.

The elegant phalaropes spend most of their time in the sea, even at high tide, feeding among surface plankton.

species perfectly capable of coping with such problems and thereby profiting from the nutritious protein contained inside the solid, tightly sealed shells. The oystercatchers provide the most striking example of this specialised technique. In their case the long, powerful, flattened bill serves as a kind of lever or crowbar, specifically designed, so it would seem, for procuring this type of food. When they come across a mollusc with the two shell sections partially open, they insert their beak into the crack and, with a vigorous toss of the head, force the animal to open up completely by severing the adductor muscles.

Oystercatchers therefore occupy the same habitats as ringed plovers, curlews and godwits without directly competing with such species. The avocets and stilts have likewise solved the problem of coexistence by fishing in the shallows where the water comes up over their legs to the level of the belly. Thanks to the length of the tarsi and metatarsi, such birds can wade into the sea to a depth of 8–12 inches without a risk of wetting their feathers. They have the added advantage of possessing webbed feet, which means that they can continue fishing even if momentarily swept off balance, despite the fact that they are not true swimmers.

Among waders with somewhat less specialised hunting methods are the godwits and redshanks, with long legs and moderate-sized beak, related to the woodcocks and snipes. Sometimes they probe for food in the mud, like the plovers, but often they emulate the avocets and stilts by venturing into the sea for a short distance. The curlews too, with their distinctively curved bill, find their food in the sand, in rock clefts and among seaweed.

In addition to these major groups of wading birds which seek food on the seashore there are species that display intermediate habits, a few of which exhibit somewhat unusual forms of specialisation. One curious example is that of the wrybill plover (*Anarhynchus frontalis*) of New Zealand, whose mandibles are shaped in such a way that it can only probe among shingle in a rightward direction. The turnstone too has the strange habit, as its name suggests, of rummaging among stones and pebbles for its small animal victims; and the spoonbilled sandpiper (*Eurynorhynchus pygmeus*) feeds by filtering mud in a similar manner to the shoveler.

The graceful phalaropes, with lobate webs on their toes, spend a large part of their life in the water, swimming around and stirring up miniature whirlpools with their feet to stimulate the movement of tiny plankton, which, becoming more easily visible, are promptly scooped up.

Another interesting and unusual manner of avoiding competition with other marine species is provided by the collared pratincole, a bird with long, slender wings, forked tail, weak legs and a short bill with enormously wide gape for catching insects in flight—an example of adaptive convergence with swallows and swifts. By feeding principally on insects, the collared pratincole can coexist with other Charadriiformes, reserving for itself a vacant ecological niche.

Gulls and terns

As on the seashore, the food resources of the coastal waters are well distributed. The most familiar birds of these shallow zones are the gulls (with an enormously varied food range) and the elegant terns or sea swallows—including black terns—which seize small fishes and crustaceans swimming at or near the surface in a pincer-like grip.

The gulls, sturdy relatives of the waders, have webbed feet to facilitate swimming and well developed wings for flying. Even on land they walk with surprising grace, both over rocks and sand. When it comes to feeding they are ready to pounce on anything that offers, live or dead—scooping molluscs out of rock cavities or swooping on carrion thrown up by the tide. They will glide over the water, on the lookout for shoals of fish, diving down on their prey, and will often pillage the nests of other sea birds, eating both eggs and chicks. Thus they are content to feed either on land, on the sea surface or in the air, frequently chasing other gulls and forcing them to regurgitate or loose their hold on recently captured prey.

The related terns, tireless fliers, avoid competition with

Herring gulls gather in large flocks around seacoasts. Opportunists, with enormous appetites, the birds are often seen in harbours, swooping down on any kind of rubbish. They will attack the nests of other birds and often squabble among themselves for prey.

Cormorants nest on cliffs, often in company with gulls and terns. They are recognisable by their black plumage, their long, sinuous neck and their short, partially webbed feet. The Shag, shown here, dives to a depth of more than 60 feet.

the gulls by being more specialised in diving for their prey and by hunting almost exclusively in the shallows. But gulls and terns differ too in the type of food consumed. Whereas the former concentrate on fairly large prey, as well as carrion, the latter usually confine their activities to smaller, more rapidly-swimming fishes or, as in the case of black terns, to tiny organisms such as plankton.

Other surface hunters are the black skimmers of the genus *Rynchops*, but they too, as a result of an individual style of fishing, avoid clashing with gulls and terns. They literally rake the water surface and snatch fishes with the aid of the lower mandible, markedly larger than the upper one.

Hunters of the deep

For those sea birds which roam farther afield into deeper waters, here too there is enough food for all. The diverse species—guillemots, puffins, razorbills, cormorants and sea ducks—not only fish at different depths but also have their individual preferences as far as habitats and type of prey are concerned. Thus while certain birds, including the guillemots, catch fishes that are normally found in shoals, others feed on the more solitary species that seek shelter among algae, as

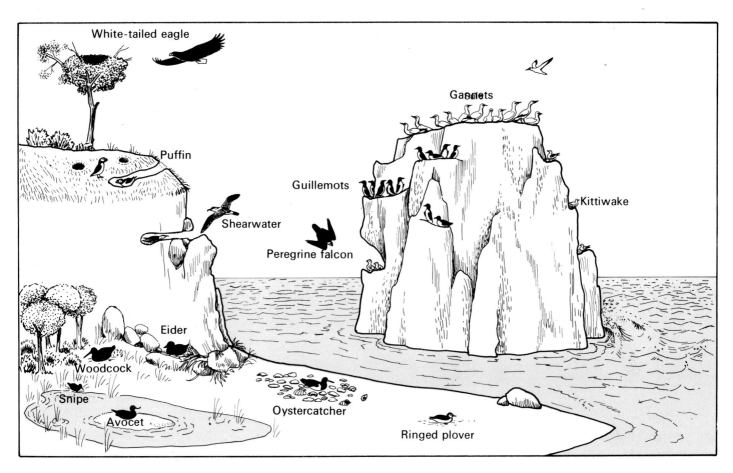

This diagram shows the many different sites that may be chosen by coastal sea birds for nesting, either on rocks, in large colonies, or alone in burrows, among stones or on the bare ground.

well as extending their range to echinoderms and crustaceans. Certain diving ducks, such as scoters and eiders, eat algae and various crustaceans living on the rocky or sandy sea bed.

Cormorants belonging to several species occupy particular habitats. The shag, for example, feeds on fishes such as wrasses which are generally found among stones and rocks, whereas the common cormorant prefers either to dive for species normally living on sand, such as flatfishes and eels, or hunts in the open sea.

It is interesting to note that one type of duck, the merganser, with its long, sharply hooked bill, feeds on fishes in much the same manner as cormorants, distinguishing itself in this respect from other diving ducks.

Breeding problems

In spring, when the breeding season commences, the sea birds of the coastal belt, in choosing a site for their nests, undergo a strange reversal of behaviour in comparison with their winter life pattern. The waders, for example, which have previously assembled in flocks, now become solitary. The gregarious instinct suddenly disappears as birds in couples disperse in all directions, searching for a suitable nesting site among the sand dunes, in clefts of rocks, in tall reeds or even, in some cases, on the bare ground. At the same time bright patches of colour appear on their hitherto fairly nondescript plumage, which has served to merge with their surroundings. All their social tendencies now give way to territorial instincts. Each male prepares to defend his little domain fiercely. This frequently results in ritual confrontations, as seen, for example, among rival ruffs.

On the other hand, species which have led a more or less independent life during the winter, perhaps forming small, temporary groups, now display gregarious attitudes, coming together to nest in enormous colonies that number thousands of birds. Such assemblies are both interspecific and intraspecific, consisting either of different species or the same species. A single line of cliffs, for example, may serve as nesting site for many thousands of gulls, guillemots, razorbills, shags, cormorants and puffins; but each species will concentrate in the particular place where it can most conveniently find food. In the North Atlantic such birds may be joined by boobies, petrels and shearwaters, filling every crag and ledge.

The division of territory among these various species is no matter of chance but stems directly from their reproductive needs and habits. Kittiwakes and shags, which construct proper nests, choose rock ledges for the purpose. So do guillemots and razorbills, whose nests have no protective walls apart from a few pieces of dry seaweed. Whenever the incubating parents enter or leave the nest, the pear-shaped eggs revolve on the narrow end so that there is no risk of their tumbling into space.

Puffins frequently build their nests in the abandoned burrows of rabbits or in hollows which they themselves scoop out. Some puffins make use of deeper cavities in the rocks.

Boobies usually take over the widest and flattest portions of a cliff ledge and if need be will dislodge other birds that happen to be there.

This nesting pattern is made more complicated by the arrival of other species such as choughs, jackdaws, peregrine falcons and, occasionally, Eleanora's falcons, all of which sometimes nest on cliffs.

The seashore raptors

With such a teeming population of birdlife on and around the coasts, it is not surprising that predators should be attracted. Worthy of special mention are the powerful sea eagles, birds of prey with a varied diet and extensive hunting grounds. Then too there are many scavengers and opportunists to ravage the breeding colonies. Besides gulls, skuas and crows, active predators in these areas include kites and marsh hawks.

As previously mentioned, there are two species of falcon which habitually nest along the seashore and which may be regarded as typical hunters of other birds. Various communities of peregrine falcons appear to specialise in attacking sea birds, and to this end live almost exclusively on certain islands, such as the Volcano Islands in the Pacific Ocean, as well as along rocky coastlines, as, for example, in British Columbia. These falcons, noted for their amazing high-speed dives from great heights, pose no threat to the closely packed breeding colonies on rocks. Their choice victims are individual birds

The kittiwake, unlike other gulls, often strays far out to sea. During the autumn and winter flocks will hunt shoals of fishes and follow fishing vessels on the high seas. In spring they return to breed on coastal cliffs, often with guillemots.

Guillemots gather in their tens of thousands on cliff ledges to breed but do not build proper nests. The pear-shaped eggs balance in such a way that they do not often fall, even though the narrow nesting sites are often overcrowded.

that may have strayed far from the nesting site and these will be attacked and killed in mid-air. Because of this hunting pattern the falcons set off each morning for the open sea, on the watch for birds on the way to and from their nests.

Marine birds are not the only potential victims of peregrine falcons, for many land birds, in the course of their migration journeys, are equally acceptable. Any such bird surprised by a falcon too far from the ground to seek refuge is virtually lost, for if not captured at the first attempt it will almost certainly be caught after a brief, desperate chase.

The gyrfalcons of the Arctic, also nesting on the coasts, perform a parallel role to the peregrine falcons farther south, the principal difference being that they are capable of tackling larger prey.

Eleanora's falcons exhibit a remarkable form of adaptation. Wintering in Madagascar, the birds return to breed on the rocky coasts and islands of the Mediterranean and on the Atlantic shores of North Africa. Because they breed fairly late, their appearance does not coincide with the spring migrations of Passeriformes (as does that of peregrine falcons) but with their outward journey in the autumn to winter quarters. Thanks to this time-lag the raptors can rear their chicks on cliff ledges without risk of confrontation with breeding peregrine falcons. But since there are few small islands and cliffs along the migration routes, Eleanora's falcons are compelled to nest in colonies, unusual behaviour for bird-hunting raptors which are normally territorial and solitary by habit.

Some representative species of Charadriiformes.

CHARADRIIFORMES

WADERS

Jacanidae	Jacana (1)
Rostratulidae	Painted snipe (2)
Haematopodidae	Oystercatcher (3)
Charadriidae	Lapwing (4) Golden plover (5) Ringed plover (6)
Scolopacidae	Woodcock (7) Curlew (8)
Recurvirostridae	Avocet (9)
Phalaropodidae	Grey phalarope (10)
Dromadidae	Crab plover (11)
Burhinidae	Stone curlew (12)
Glareolidae	Cream-coloured courser (13) Collared pratincole (14)
Thinocoridae	Seedsnipe (15)
Chionididae	Sheathbill (16)

GULLS

Stercorariidae	Skua (17)
Laridae	Herring gull + Blackheaded gull (18) Tern (19)
Rynchopidae	Black skimmer (20)

AUKS

Alcidae	Guillemot (21) Puffin (22) Razorbill (23) Little auk (24)

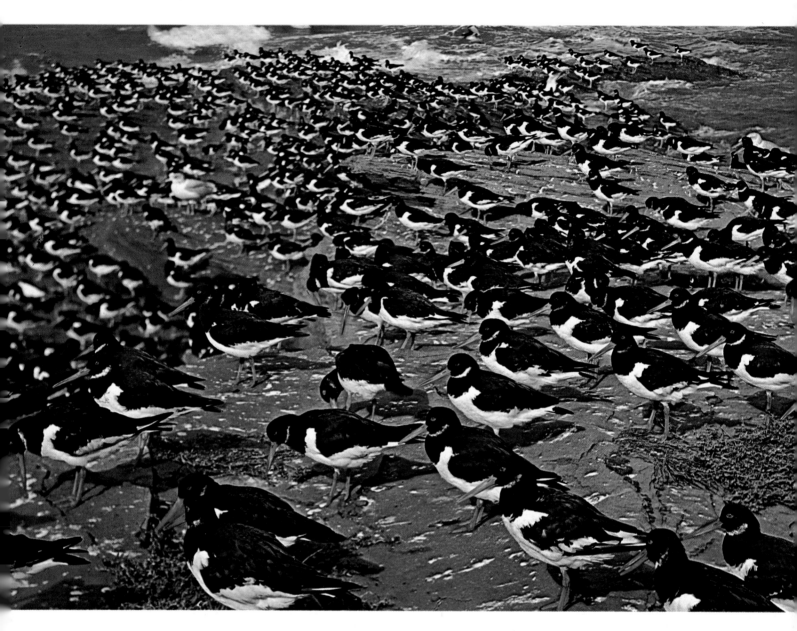

The versatile Charadriiformes

The order Charadriiformes groups together an immense crowd of birds which differ radically from one another both in appearance and structure. Because the group is so enormously complex, ornithologists have been obliged to create a large number of subdivisions, each exhibiting a variety of adaptations.

The birds properly described as waders belong to the suborder Charadrii, which comprises two particularly large families that contain the majority of species. These are the Charadriidae (plovers and lapwings) and Scolopacidae (sandpipers, snipes, curlews, godwits, turnstones and the like).

The suborder as a whole brings together birds which habitually haunt shores and beaches. They generally tend to live close to the water's edge, whether this be fresh or salty, although there are certain species which are confined to desert zones. Small or medium-sized, these waders normally spend their time on land and there too, for the most part, they build their nests and breed. One common characteristic is that they are chiefly found on flat, exposed terrain, protected as they are

Oystercatchers, like other waders, gather in large flocks on the seashore, their variegated plumage making a picturesque effect. They are easily identified by their pink legs and orange bill.

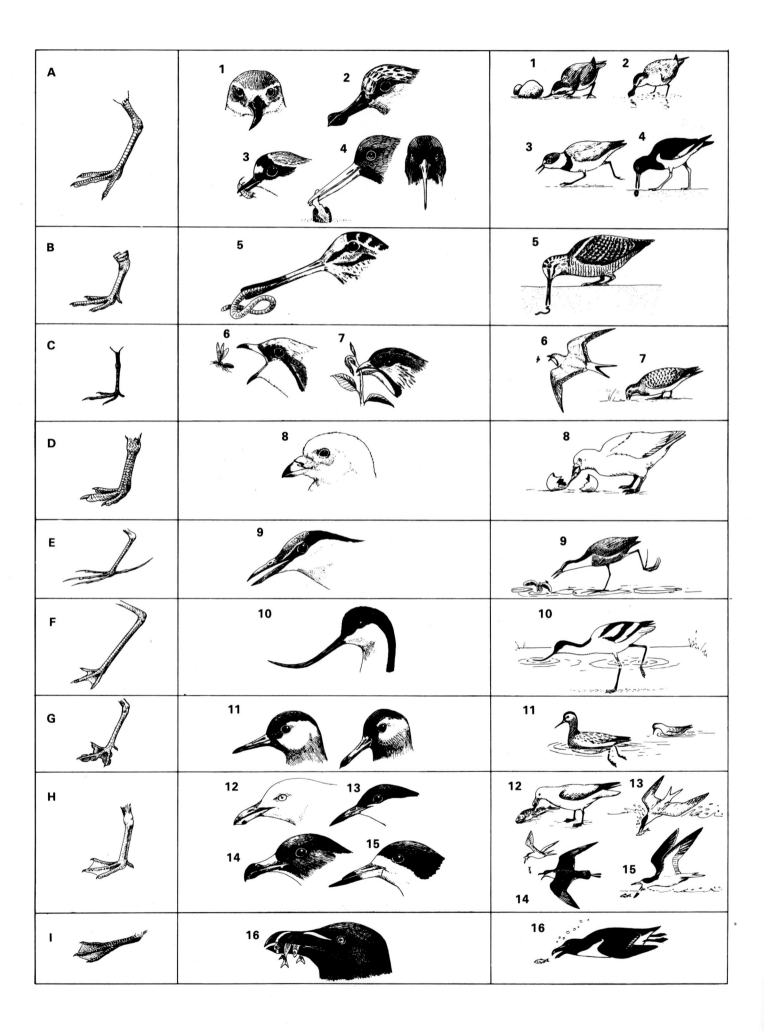

either by the cryptic coloration of their plumage or the ability to fly rapidly away if danger threatens. Exceptions to the rule are the woodcocks which prefer tree cover and the snipes which normally seek refuge among reeds. Many species of this varied suborder possess plumage with contrasting bright marks, evidently with a view to allowing individuals to recognise and communicate with one another.

The Charadrii are excellent runners and have the habit of scampering swiftly along the ground for a few yards before actually taking wing. In the air their flight is powerful, fast and sustained. Many of them are migrating species, including certain birds which, hatching on the tundras of Canada and Eurasia, fly off, at a comparatively tender age, to pass the winter in South America or equatorial Africa. The representatives of the group are for the most part diurnal, but a few, such as stone curlews and woodcocks, are active by night, as is clearly revealed by their eyes.

There is also astonishing diversity in the shape of the bill. Generally such modifications are responses to food needs, which are almost always of animal origin. There is also a direct relationship between the size of the bill and the length of the legs. The toes may be separate, webbed, partially webbed or lobate. The hind toe, when not absent, is often very short. As for the wings, they are frequently composed of eleven remiges and sharply pointed.

Except during the breeding season, these wading birds customarily congregate in large flocks which sometimes number more than a thousand individuals.

The Charadriidae and Scolopacidae have a wide geographical distribution, being represented in all parts of the world. But there are also other families in the suborder Charadrii, namely Recurvirostridae (stilts and avocets), Phalaropodidae (phalaropes), Haematopodidae (oystercatchers), Jacanidae (jacanas), Rostratulidae (painted snipes), Chionididae (sheathbills), Glareolidae (coursers and pratincoles), Dromadidae (crab plovers), Thinocoridae (seedsnipes), and Burhinidae (stone curlews). But these contain fewer species and are not so wide-ranging.

Oystercatchers

The oystercatchers of the family Haematopodidae are sturdy, medium-sized waders. The plumage of these tireless, rather noisy birds is usually black above and white below. The feet and legs are pink and the blunt, flattened bill is orange, the tail is short and rounded. But there are striking variations of colour, both sexual and seasonal. The plumage of young oystercatchers tends to be brown, only later acquiring the glossy black sheen of the adults, and their feet and bill are paler in colour.

Normally active during the day, oystercatchers often let out loud cries at night. They are found in river estuaries, coastal lagoons and beaches, whether sandy or rocky. They are agile runners, excellent fliers and, when necessary good swimmers.

Outside the breeding season oystercatchers form large flocks containing hundreds of birds. But when the time comes to breed,

Facing page: The Charadriiformes have adapted to conditions on and around the seacoasts in many individual ways. Differences occur mainly in the structure of the feet and bill, according to their feeding habits, as shown in these drawings. The species in group A are rapid runners, those in B and D normally walk over sand and shingle, and those of group C generally capture insects on the wing. Birds of group E feed on semi-submerged plants and those in F are intermediate forms, adapted both to dry land and shallow water. The species shown under G, H and I are among many that swim and dive. The birds illustrated are: 1. Wrybill plover. 2. Spoonbill sandpiper. 3. Wilson's plover. 4. Common oystercatcher. 5. Woodcock. 6. Collared pratincole. 7. Seedsnipe. 8. Sheathbill. 9. Jacana 10. Avocet. 11. Red phalarope. 12. Herring gull. 13. Arctic tern. 14. Long-tailed skua. 15. Black skimmer. 16. Razorbill.

Wrybill plover
(*Anarhynchus frontalis*)

Spoonbill sandpiper
(*Eurynorhynchus pygmeus*)

The oddly-shaped bills of the wrybill plover of New Zealand and of the spoonbill sandpiper, an inhabitant of north-eastern Siberia, are adapted to extracting small animals sheltering under rocks or burrowing into mud and sand.

the adults pair off, often leaving young, immature birds on their own. Spring sees diverse courtship rituals, the birds strutting around with outstretched neck and downward-pointing bill, and a great deal of noisy chattering.

The bulky nest is placed in a shallow depression in the ground, where the female lays two to four eggs. They are speckled with dark brown or black marks and the shells are thus perfectly camouflaged. Both parents cooperate in incubation, taking it in turns. The newly hatched chicks are well developed and are fed by the parents for about a week.

Lapwings and plovers

The Charadriidae are medium-sized or small birds with a large, rounded head and compact body. The bill, which may be short or of moderate length, is narrow and swollen at the tip. The eyes are fairly large. The long, pointed wings are powerful, the tail usually small. The legs are of variable length and the feet furnished with three or four shortish toes. The plumage, varying according to species, exhibits a range of colours which blend into all manner of patterns and change with the seasons, although normally blending with the surroundings. The birds often display a white stripe on the nape and there are dark bands on the thorax and fringing the tip of the tail. All species are gregarious, except while breeding.

There are about 60 species in this family which are distributed all over the world. But although found in the northern hemisphere, they are even more abundant in the tropics. Individuals breeding in the more northerly latitudes have migratory habits.

Three principal groups, easily distinguishable, make up the family. The lapwings are identified by their white, black-tipped tail and their black primary feathers, adorned by a white band. Additionally, almost all lapwings have a characteristic crest, such as no other Charadriidae possess. A number of them display a spur on the angle of the wings. The Ethiopian region is the principal home of lapwings.

The second group comprises the half dozen or so species of plovers belonging to the genus *Pluvialis,* including the golden and grey plovers. The plumage of the back is richly and variously coloured, and during the breeding season there may be a black mark on the abdomen.

In the third group, the twenty odd species of plovers of the genus *Charadrius* (including the ringed plovers and the sand plovers) are fairly small birds with brown or grey plumage above, white below, and a characteristic pattern of black bands on back and breast, between which are other prominent white stripes.

Apart from the afore-mentioned principal groups, there are some outsiders, including the odd-looking wrybill plover (*Anarhynchus frontalis*) of New Zealand, with a long, asymmetrical bill which bends to the right, and the little known Mitchell plover (*Phegornis mitchellii*). This silent, rather solitary bird is an inhabitant of South America, frequenting the rocky banks of mountain streams and rivers. Its alternative name is diademed sandpiper plover.

Sandpipers, curlews and relations

The family Scolopacidae, very large and very varied, includes such dissimilar groups as sandpipers, ruffs, redshanks, sanderlings, curlews, woodcocks and snipes.

All these birds live close to water. Like the Charadriidae, except during the breeding season in the spring, they form fairly large bands which perform beautifully synchronised aerial manoeuvres in which the contrasting black backs and white underparts are clearly visible.

The majority of these species prefer to frequent open ground near water, although the woodcocks and snipes remain all the year round in forests and marshes respectively. Some species tend to be most active at dusk or are overtly nocturnal. The song pattern is naturally extremely varied, ranging, according to species, from strident cries to modest chirps. One group of snipes even produce mechanical, non-vocal sounds as a result of air friction against the tail feathers. Such sounds, whether vocal or mechanical, are wholly or most commonly associated with courtship display.

Even if an oystercatcher is disturbed in the course of incubation, her eggs are difficult for an intruder to locate, so perfectly do the colours match the surrounding rocks and stones.

Reproductive habits too are highly variable. The majority of Scolopacidae nest on open ground or among reeds, although some take over the empty nests of other species. The eggs, generally four in number, blend with the surroundings.

Most of the birds of this family, comprising approximately 70 species grouped in 24 genera, are inhabitants of the northern hemisphere and many species have a circumpolar distribution. These birds undertake extensive migration journeys, principally because their particular food habits make them vulnerable to winter frost and ice, making it impossible for them to probe deeply into the rock-hard ground.

The woodcocks and snipes, seldom if ever seen on the beaches, are distinguished by their long powerful bill, the upper part of which has a flexible point.

The godwits and sandpipers (the latter also including ruffs, sanderlings, knots and dunlins) make up yet another group.

Like many waders, the common sandpiper makes her nest on the ground, the grey-brown plumage blending well with seashore vegetation.

The phalaropes

The three known species of phalaropes form a family of their own (Phalaropodidae), probably having broken away from others of the order in very ancient times. Among a number of peculiarities exhibited by these birds is the fact that the male is smaller than the female and that, surprisingly, she possesses more elaborate nuptial plumage.

Phalaropes are pelagic birds which spend most of their time swimming. The marked lateral flattening of the tarsi and the lobed membrane of the toes are signs of these aquatic propensities. Furthermore the long, thick plumage retains quantities of air and this facilitates floating.

Bobbing about on the crests of the waves like little paper boats, these graceful wading birds jerk their heads back and forth as they propel themselves along, often turning in small circles as they peck at tiny organisms on the surface.

Phalaropes are to be seen on lakes, ponds and pools in the higher latitudes of the northern hemisphere, with a circumpolar distribution. Wilson's phalarope usually breeds in the western parts of the United States and Canada, but there is an independent group in the eastern Great Lakes region.

The dove-like sheathbills

Possessing certain features common both to waders and gulls, the sheathbills live in the Antarctic regions. Although unrelated to Columbiformes they bear a strong resemblance to white doves and are in fact referred to in South America as sea doves. The family Chionididae comprises two species, the wattled sheathbill (*Chionis alba*) and the lesser sheathbill (*Chionis minor*), both with a shield-like, horny sheath over the base of the upper mandible and under the eyes, from which they derive their common name.

Sheathbills are for the most part land birds, searching incessantly for food on rocks and shingle. The main constituents of their diet are crustaceans, caught among the enormous clumps of seaweed thrown up on the shores by ocean currents. But their food range is, in fact, far more extensive. The birds are both scavengers, swooping on any kind of carrion or detritus, and also predators, attacking colonies of sea birds for eggs and chicks. They occupy a similar ecological niche to that of gulls and crows. Sheathbills have frequently been observed assaulting adult penguins at the very moment that the latter prepare to feed their chicks, forcing the terrified birds to regurgitate the fish they have caught.

During the breeding season sheathbills assemble in small groups. The females lay two or four eggs but it would appear that often only one chick survives after an incubation of 29 days.

With the arrival of spring the sheathbill communities along the frozen Antarctic shores make their way towards the less rigorous northern zones of their range. Populations of these subantarctic regions appear to have sedentary habits and do not make even limited migrations.

With its long legs and slender sword-like bill, the black-tailed godwit (*above*) fishes in muddy shallow water. The wattled sheathbill (*below*) feeds on crustaceans, eggs and chicks of other sea birds, and carrion.

Franklin's gulls (*above*), like related species, are superb fliers, and enormous flocks travel considerable distances from roosts to fishing grounds and back. The drawings below show how the various groups of Charadriiformes have gradually adapted to marine conditions. Waders such as the ringed plover (1), with a light skeleton, are excellent runners and fliers. Sea birds such as the puffin (3), with heavy bones, specialise in swimming. In between come the gulls (2).

Gulls and terns

The suborder Lari comprises yet another major group of the Charadriiformes and is made up of three families. The Laridae include the many species of gulls and terns. The Stercorariidae are the four species of skuas (rapacious birds which are notable for their habit of indefatigably chasing other sea birds until the latter regurgitate and drop prey they have recently swallowed). The Rynchopidae are the skimmers, birds that swoop low over the sea surface, scooping fish out of the water with their unusually long lower bill.

The best known of these birds, because the most cosmopolitan in range, are the gulls, familiar to all who live on and around the seacoast. But although most of the Laridae frequent coastal waters, there are many which are in the habit of roaming farther inland so that they are frequently seen swooping and wheeling over rivers, lakes and marshes.

Most of the gulls are fairly large, the body strong and thickset, the tail short, the wings long and pointed. The beak is powerful and slightly hooked. The three front toes are joined by a web and the rear toe is very small.

Although there is no essential difference in colour between male and female, the shades and patterns of the plumage vary widely according to the bird's age. Thus whereas an adult may have almost plain white or pale grey plumage, the characteristic colours of a young gull's feathers are greyish-brown with irregular darker patches. Among adult birds too there will be variations based on age and time of year. In the breeding season, for example, many species of gulls take on a form of black mask which covers the front part of the head. As a rule the bill and feet of the adults are brightly coloured—a vivid red, orange or yellow.

A typical gull's nest is an unadorned heap of grass or dry seaweed. According to species, it will be situated either on the cliffs, in marshes among reeds, on sand dunes, on the water surface or occasionally in a tree. The female customarily lays two or three eggs, the base colour being brown or greenish and the shell heavily spotted with darker marks. Both parents take turns in incubation and later join in feeding the nidifugous chicks.

Thanks to their pattern of rapid and sustained flight, gulls are capable of covering considerable distances and some species

The celebrated naturalist Niko Tinbergen, in his study of the behaviour of herring gulls, distinguished various attitudes adopted during the breeding period. The posture of the male (1) deters rivals and also attracts the female. Should an intruder fly too close to his territory he will extend his neck and emit loud repetitive cries (2). Having formed pairs, male and female take up a position side by side (3), then turn their heads away from each other (4), concealing the facial parts which normally stimulate the aggressive tendencies of sexually excited males (5). Eventually the birds turn towards each other and the female adopts an infantile attitude, inviting the male to feed her (6). The courtship ritual is followed by coupling and egg-laying.

Niko Tinbergen discovered that when a baby herring gull is hungry it forces its mother to regurgitate food by pecking against the red mark on her lower mandible. Tinbergen proved that the sight of this mark always unleashed the same reaction in the chick.

Preceding pages: Newly hatched Caspian terns have an enormous appetite and devour large fishes caught by the parents at an amazing rate.

undertake long migrations. Although the birds are often seen floating on the water they are not, in fact, noted for their swimming powers. The large amount of air trapped in the long, oily plumage is both an asset and a handicap, for although it enables them to bob about comfortably it offers too much wind resistance. Apart from this, swimming capacity is diminished by the fact that the body is not sufficiently streamlined and that the muscles of the legs are relatively weak.

The diet of these remarkably opportunistic birds is extremely varied. Although some species group together for fishing expeditions, the majority, whilst including fish in their diet, are scavengers, preferring to feed on dead fishes and indeed any other available type of refuse, whether of animal or vegetable origin. It is hardly surprising, therefore, that with such varied habits gulls should congregate in areas where the diverse activities of man afford food virtually for the taking, especially in ports and harbours. In some coastal regions of the North Sea their numbers have multiplied to such an extent that it has been necessary to take drastic steps for controlling their population, because of the serious threat posed to other sea and marshland birds in the vicinity, whose nests are systematically pillaged and whose chicks are killed.

According to M. Moynihan, who has made comprehensive surveys of gull behaviour, it is convenient to divide the birds into three natural groups. Firstly, there are the white-headed gulls, such as the glaucous gulls, herring gulls, common gulls, lesser and great black-backed gulls, etc; secondly, the black-headed gulls, including the species of that name, the laughing gulls, little gulls, Mediterranean gulls, etc (these being smaller than the previous group); and thirdly, on its own, the ivory gull, which breeds in the Arctic, is completely white and thus differs from the preceding species.

The terns or sea swallows, although related to the gulls, are usually much smaller. They are distinguished too by their forked tail, straight pointed bill, weaker legs, virtually unwebbed feet and longer, narrower wings. Moreover most of their food consists of live prey such as small fishes and crustaceans, caught by diving from a height into the water, in the same manner as gannets and boobies. Marvellously adapted to long-distance flight, certain species, such as the Arctic tern, are famous for their migrations which in this case take the birds more than 20,000 miles away from and back to the breeding grounds.

The water-loving alcids

The family Alcidae—auks, puffins, razorbills, guillemots, etc—are grouped together in the suborder Alcae and are indeed so distinctive that in the view of some authors they are entitled to form a separate order (Alciformes). In their many habitats these birds are better adapted to conditions in the sea than any other major group.

The body is long and heavy, the feet are webbed and the legs are placed well to the rear, all of which makes them excellent divers and swimmers. Yet so perfectly suited are they to their

After an elaborate courtship ritual, laughing gulls generally couple on the nest they have just built.

watery environment, in fact, that they do not swim under water, like the majority of the world's diving birds, by paddling their feet, but propel themselves along with their powerful little wings, so that they are literally flying below the surface. The feet, outstretched behind them, only come into play as rudders when the birds change direction.

The Alcidae live in the colder regions of the northern hemisphere but their range extends southwards to the temperate waters of the Gulf of California. The ancestors of this group are believed to have separated from waders and gulls over 70 million years ago. Today there are 23 species in 14 genera.

There is much variation in size among this family. Thus the little auk is no larger than a starling (about 8 inches), whereas the now-extinct great auk measured up to 32 inches. The colours of the plumage, predominantly white, grey and black, are arranged in such a way that the dark tones tend to cover the back and the paler shades the belly. This colour pattern helps to camouflage the bird in the water, concealing it effectively from predators both in the water and those in the air. Some species, such as the tufted puffin, have long crests of feathers on either side of the neck. Frequently the naked areas of the body, including the bill, take on brilliant hues during the breeding season, as is the case with the common puffin. More commonly a white pattern develops around the eyes or between the eyes and bill, somewhat resembling a bridle or a pair of spectacles.

The tail is always short but the form varies, according to species, being either rounded or pointed. The colour and pattern of the plumage change with the seasons. Another feature of

Certain gulls are so obstinately attached to their nesting site that if the eggs are removed and set down only a few feet away they will ignore them and resume incubation on the empty nest.

the birds is that their wing feathers all moult at the same time so for part of the year they are incapable of flying.

All these birds spend the greater part of their time in the sea and only venture onto land for nesting purposes. Should they be disturbed they automatically dive into the water and it is only in very exceptional circumstances that they will take wing. This is never an ideal solution for them, particularly when they are gorged with fish. Even when they have not fed recently and have the advantage of a following wind, they only manage to raise themselves into the air with considerable effort and a series of heavy, exhausting wing flappings. Once aloft they are obliged to make up for the absence of a large supporting surface by continually beating their small wings, and the noise they make in so doing is reminiscent of the buzzing of myriads of bees.

The majority of species nest in huge colonies on cliffs and rocks, among seashore vegetation or even in marshes. The rough nest is built with only a few scraps of plant matter. As a rule the females lay one or two eggs. Those of guillemots and certain other species which make their nests on high ledges are pear-shaped, so that they automatically revolve on their narrow end and do not roll off. Many of the chicks, including those of guillemots, are extraordinarily precocious, hurling themselves from rocks into the water, in the wake of their parents, only a few days after they are born. But others, such as common puffins, remain in the nest for about seven weeks.

Brünnich's guillemots, like related species, are among the best adapted to marine life, swimming in deep water by using their wings as paddles.

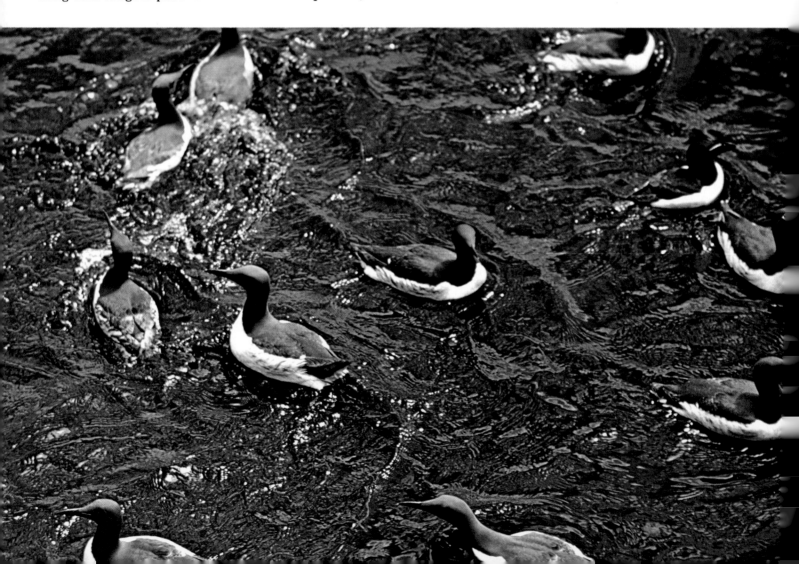

These various sea birds tend to be silent except while rearing their young, when they give vent to raucous cries.

Some of the more thoughtless fishermen of North Atlantic seacoasts collect the eggs of guillemots and razorbills, not hesitating to kill adult birds while they are incubating. The widespread harm caused by such activities has resulted in the establishment of sanctuaries (as in North America, the British Isles and the Baltic countries) where the larger colonies can nest undisturbed, in guaranteed security.

The great auk became extinct for this very reason. The species was particularly well adapted to ocean life (for although it possessed flight feathers on its wings it was incapable of flying). Its numbers had already diminished rapidly as a result of a series of events that proved unfavourable for breeding, including upheavals on the volcanic islands of the North Atlantic where its principal colonies were situated. Man completed the catastrophe. The last recorded pair of breeding birds were sighted on a cliff by the crew of a ship, who promptly killed the female and removed the single egg – the sad climax to many similar acts of stupidity and barbarity.

The lordly sea eagles

As a violent storm descends on the coasts of the North Sea, an unnatural calm prevails over land and sea. The gulls hang almost motionless in the air, the plovers and other birds of the seashore shelter from the wind among the dunes, the cormorants huddle on their rock ledges. Even the waves themselves, grey and flecked with foam, seem to be of stone, massive, menacing and unfriendly.

The raging of the elements has little effect, however, on the predatory activities of the majestic sea eagles, gliding calmly above the turbulent waters on their huge wings, apparently suspended in space as they survey their hunting grounds below. All the animals of these coastal regions, whether fish, mammal or bird, have good cause to fear these mighty raptors.

The sea eagles belong to the single genus *Haliaeetus*, included among the Accipitridae. The eight species are Sanford's sea eagle (also known as the Solomon Island's sea eagle); the white-bellied sea eagle, with a range extending from Ceylon to South China and Tasmania, by way of New Guinea and the Bismarck Archipelago; the African fish eagle or river eagle, found in the Ethiopian region; the Madagascar fish eagle, confined to that island; Pallas' sea eagle, an inhabitant of central Asia; the white-tailed sea eagle, found in Europe but also ranging as far afield as Manchuria, the Kurile Islands and Hokkaido; the bald eagle, national emblem of the United States; and finally, the largest, strongest and most beautiful member of the group, Steller's sea eagle, of continental Asia. Since it is the white-tailed sea eagle which has been most thoroughly studied by European ornithologists, our description here will be confined to that species.

The pure white tail of this bird stands out in vivid contrast to the brown colour of the rest of the plumage. Bill, iris and

Recognisable by their flattened beak and thin white markings on head and wings, razorbills often nest in colonies on rock ledges along the seacoasts.

feet are yellow. The young bird has a steel-grey beak, chestnut iris and yellowish feet and cere, but the body feathers, including the tail, are brown. The seashore is the favourite habitat of the species although it occasionally ventures into the valleys of larger rivers or seeks shelter in lakes and on islands. Pairs of birds generally remain all the year round on their own territory, which may vary from 15 to 75 square miles. Only in the course of exceptionally harsh winters, with severe frost and ice, will the normally sedentary birds make partial migrations. But even when the birds fly over their territory in quest of food they appear to spend most of their time hovering lazily in the air, conserving energy.

Like other related species, the white-tailed sea eagle is extremely large. The wingspan of the adult male may exceed 7 feet and that of the female slightly more, around the 8-foot mark. The vocal range includes a wide variety of throaty cries but these seem comparatively feeble in relation to the dimensions of the bird.

The eagle has a wide-ranging diet. Fishes are caught by means of short but accurate dives (one of the favourite victims being

In the breeding season brilliant orange growths develop on the bills of common puffins, contrasting strikingly with their black and white plumage.

the pike), but the technique is less spectacular than that of the osprey. It has also been observed hovering on the brink of shallow pools, probably with a view to taking fishes unawares. In addition the eagle captures large numbers of water birds, especially the slow-moving coots. To do this it flies low, almost raking the surface, launching surprise attacks on its victims. Eiders and other diving birds are often chased to the point of exhaustion, for despite the enormous size of the raptor and its apparently lazy gliding and hovering attitudes, it will, should the need arise, pursue its prey relentlessly. The species that elude the sea eagle, however, are swift fliers such as gulls and terns, although it will kill their young if given the chance.

Mammals also feature occasionally in the white-tailed sea eagle's diet. Although rats and fawns are sometimes taken, choice victims are rabbits and hares. But according to the observations of Brown and Amadon, conducted during the breeding season, birds make up 50-60 per cent, and fishes 25-30 per cent, of the raptor's diet. Carrion is also consumed, and each bird is reckoned to eat a little over one pound of food daily.

The average weight for a male is 9 lb, for a female 11 lb.

Although many sea birds of the family Alcidae are black above and white below, the black guillemot is distinctive for being almost completely black, intraspecific recognition being assured by the white wing patch.

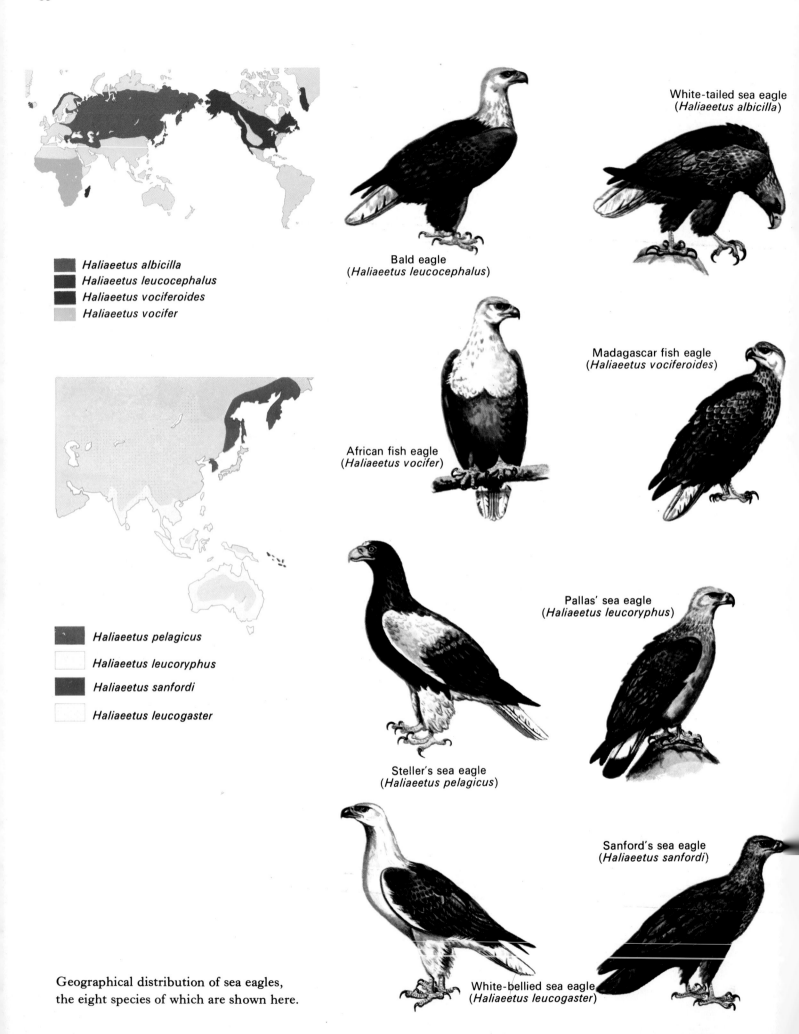

Geographical distribution of sea eagles, the eight species of which are shown here.

Haliaeetus albicilla
Haliaeetus leucocephalus
Haliaeetus vociferoides
Haliaeetus vocifer

Haliaeetus pelagicus
Haliaeetus leucoryphus
Haliaeetus sanfordi
Haliaeetus leucogaster

Bald eagle (*Haliaeetus leucocephalus*)
White-tailed sea eagle (*Haliaeetus albicilla*)
African fish eagle (*Haliaeetus vocifer*)
Madagascar fish eagle (*Haliaeetus vociferoides*)
Steller's sea eagle (*Haliaeetus pelagicus*)
Pallas' sea eagle (*Haliaeetus leucoryphus*)
White-bellied sea eagle (*Haliaeetus leucogaster*)
Sanford's sea eagle (*Haliaeetus sanfordi*)

A pair of these eagles will stay together throughout the year and the courtship ritual may take place at virtually any season, although most often in spring, after the young have left the eyrie. In the course of the nuptial ceremonial both birds fly about together, uttering cries in unison, the male above the female, presenting his claws; but even during this aerial display it is rare for the birds to make direct contact. The major part of the courtship ritual takes place in the vicinity of the nest. Should a rival intrude upon the hunting grounds or the territory where the young are being reared, the male will expel the stranger violently, frequently after the two birds have come to blows.

Coupling occurs either in a tree or on a rock. The enormous nest is built, for preference, in the fork of a tree but should this not be possible a cliff, a hillock or even the bare ground will suffice. Frequently the same eyrie will be used year after year until it attains massive proportions. One such nest, observed in Norway, weighed over 500 lb, having been used for four successive years, by which time it contained some 2,900 branches and twigs. Both adults collaborate in the construction, sometimes adorning it with green boughs, the male carrying the materials, the female assembling them. As a rule the work takes several months but if it is a question of constructing one in a hurry, to replace one that has been destroyed, the birds may complete the work within a month.

The female lays one to three eggs (usually two), weighing about 5 ounces. The shells are plain white, sometimes speckled with yellow.

Incubation commences as soon as the first egg is laid and is generally undertaken by the female, although in certain situations the male may take her place. His principal task, however, is to provide food for his mate, although both birds may decide to hunt together during this period. Incubation lasts 36-46 days.

The newborn eaglets weigh 3-4 ounces. Their growth is relatively slow and only after a month do the first downy feathers begin to sprout. At the age of a month and a half the young start flapping their wings and are capable of tearing up their own food. By eight weeks they are exploring the nearby branches, without straying too far from the eyrie, and at ten weeks they are ready to fly. The various siblings get along well with each other and have never been seen fighting, as is the case with other young birds of prey. As time passes the parents take progressively less notice of the young and gradually the latter achieve independence. Nevertheless the adult birds continue to educate their youngsters in the rudiments of hunting. In Europe the eaglets can fend for themselves by July or August, months when there are plenty of dead animals to be found washed up on the beaches.

The various species of sea eagles reach sexual maturity only when five or six years old, so that the birth rate is extremely low. It is hardly surprising that numbers should have decreased markedly in recent years. In addition they are threatened by hunters and poachers, despite efforts to protect them and their ranges are being progressively reduced.

As a result of the uncontrolled use of insecticides, water pollution and gradual destruction of its habitats, the splendid bald eagle of North America is threatened with extinction.

CHAPTER 5

Mammals of the seashore

The first vertebrates, like the other animal groups, originated in the ocean. Only in the long course of evolution did certain individuals acquire the structures and adaptations which enabled them to breathe oxygen directly from the air and thus take up residence on dry land. First to do so were fishes, which managed to survive out of water for brief periods, but the real change to this new life style came with the amphibians. Transformation of their flippers into four limbs permitted them to walk on solid ground but they continued to be dependent upon water for reproduction and the development of their larvae. It was the reptiles which made the breakthrough by, as it were, 'inventing' the shelled egg as a reproductive method, for this enclosed, liquid-containing capsule allowed the embryo to develop away from water. The reptiles then proceeded to diversify. From them stemmed the lords of the air, birds, with forelimbs converted into wings, and the future sovereigns of dry land, mammals, carrying and feeding the embryos inside their bodies.

Yet the mammals which began to evolve some 200 million years ago did not confine their domain to the land. Some succeeded in flying, like birds, and others returned to the ocean. Many of the latter, with their streamlined bodies and flippers, bore a close resemblance to fishes and were equally at home in the sea, yet were true warm-blooded, placental mammals. One of them, still surviving, however perilously, is the blue whale, the largest animal of all time.

Today's marine mammals (among which we must include species that frequent estuaries and even roam considerable distances up large rivers) include one mustelid – the sea otter – as well as pinnipeds (seals, walruses and sea-lions), sirenians (dugongs and manatees) and cetaceans (whales, dolphins and porpoises).

Facing page: Despite its grotesque appearance, the massive walrus, with its whiskers and long tusks, is a placid, generally inoffensive, sea mammal.

Not all of these animals have adapted in equal measure to aquatic conditions. The odd man out, the sea otter, spends some of its life on land, close to the shore, whereas the sirenians and cetaceans never leave the water. In between are the pinnipeds which spend lengthy periods in the sea, even sleeping there, but which are obliged for varying periods to venture onto dry land, especially to give birth and to moult.

Apart from the problems of respiration, which these mammals have resolved by breathing air (in one manner or another) at the surface, a major difficulty for such warm-blooded animals is the maintenance of a constant body temperature. In the case of all marine mammals apart from the sea otter (whose skin is itself an insulating medium) body heat is retained by means of thick layers of fat.

The tool-using sea otter

The giant Brazilian otter, inhabitant of the great rivers of South America, weighs about 70 lb, but the sea otter (*Enhydra lutris*) is even heavier, some individuals weighing 80 lb. It is therefore the largest mustelid in the world. Its present-day distribution is, however, restricted to the coasts of California, western Alaska and certain island groups to the north of Japan, whereas at one time it ranged along all the shores of the North Pacific. The reason for its decline has been fur hunting. At the beginning of the 20th century the beautiful pelt of the sea otter was more highly valued than that of any other animal and numbers decreased so alarmingly that in 1910, when extinction seemed imminent, the United States led the way in passing protective legislation, followed in 1911 by all other interested nations. Although the species was saved, its range had been reduced by about 80 per cent.

The precious fur of the sea otter serves a vital purpose. The animal does not possess any insulating layer of fat to protect it from the cold. This indicates that not only did it populate the ocean at a relatively late stage but that it is not fully adapted to marine conditions. The long compact hairs of the pelage trap pockets of air and prevent the loss of heat. If soiled or polluted by oil, the fur, like the feathers of birds, loses its insulating qualities.

The body of the sea otter is stocky and the tail is rather shorter than that of other otters. It lives in small bands, normally in shallow waters close to rocky shores, seldom venturing on land and then never more than a few hundred yards. At night it anchors itself by wrapping a frond of large kelp around its body, often covering the eyes with its hands as if to protect them from the light of the moon. At dawn it resumes its activities. According to the American naturalist K. W. Kenyon, it dives to a depth of 200 feet for food, consisting basically of sea urchins, sea cucumbers, crabs, mussels, limpets and other molluscs. Although the animal may remain under water for four minutes, it is usually submerged only for about one and a half minutes. It catches prey with its hands and then, clutching it to the chest so that the fur forms a kind of receptacle, rises to the surface. In order

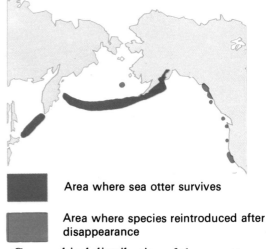

■ Area where sea otter survives

■ Area where species reintroduced after disappearance

Geographical distribution of the sea otter.

SEA OTTER
(*Enhydra lutris*)

Class: Mammalia
Order: Carnivora
Family: Mustelidae
Total length: 49–61 inches (125–155 cm)
Length of tail: 10–14½ inches (25–37 cm)
Weight: male 59½–81½ lb (27–37 kg)
female 35–66 lb (16–30 kg)
Diet: sea urchins, molluscs, fishes, crabs
Gestation: 8–9 months
Number of young: one

Sturdy body, clearly arched when the animal moves about on land. General colour dark brown, with individual variations. Large head, small neck, colour of which is usually lighter than rest of body. Short, pointed ears almost concealed under fur; small eyes. Feet larger than hands. Tail short in comparison with other otters. Baby well developed, similar to adult, with milk teeth.

Facing page: Ecological pyramid of the coastal fringe. As on land or in the open sea, all living creatures on the seacoast derive their energy basically from sunlight. The food chain starts with plants – in this case mainly microscopic algae that constitute phytoplankton – and then progresses upwards by way of invertebrates, sea fishes, birds and mammals.

The sea otter which may weigh over 80 lbs is the largest mustelid in the world.

The sea otter is one of the few animals capable of using a tool. Under the water it takes hold of a large stone to detach limpets, oysters and other molluscs from rocks (1); then, floating on the surface, it uses the same stone to hammer open the shells (2).

to prise loose a limpet from a rock to which it may be attached or to open the hard shells of any molluscs, the otter may sometimes make use of a stone. It is therefore one of the small, privileged group of mammals capable, like man, of using a tool for a given purpose.

Under water, when the sea otter finds it impossible to detach a limpet or an oyster with its bare hands, it takes hold of a stone and strikes the mollusc until it releases its grip. But it hangs on to the stone as it regains the surface, still clasping its prey. Then, flipping over onto its back, it balances the stone on its chest. Holding the prey between its forepaws it hammers it onto the stone, using the latter as an anvil. Observation of one animal showed that it required some thirty-five blows to open up a mussel but only nine for a sea urchin.

In the case of animals without shells, no such strenuous efforts are needed. The otter simply relaxes in its favourite floating position, holds the prey firmly with its forepaws and proceeds to eat it. Occasionally it catches fishes and octopuses but should these, especially the octopuses, wriggle too much, it will use a stone to stun them.

The sea otter has an enormous appetite. The amount of food eaten daily is equivalent to at least one-fifth, often one-quarter, of its body weight—between 10 and 20 lb.

When not occupied with feeding, the animal amuses itself by floating on its back, tumbling about in the waves and engaging in playful contests with other otters. It will turn over onto its belly only when forced to move rapidly, perhaps to escape a predator or hunter. As a rule it remains within half a mile or so off the seacoast, but in exceptionally stormy weather, when huge waves batter the rocks, it will seek refuge farther out to sea.

Males normally remain apart from the females, although mating may take place at any time of the year. This occurs in the water, following a fairly spectacular courtship ceremony. The animals couple by both adopting a floating position on the back, the male usually underneath. Gestation lasts 240–270 days, and the single baby is born on shore, already well developed, with eyes open, its fur similar to that of the parents and the milk teeth in evidence. Mother and baby are inseparable, with the latter clinging to her chest or abdomen while she gently floats in the water. The solicitous attention lavished on the baby by its mother is exploited by hunters who often catch a young otter so as to entice the female and kill her. Sexual maturity does not occur in the female until it is three years of age, and she gives birth only once every two or three years.

The natural enemies of the sea otter are killer whales and

Above and facing page : The sea otter, weighing up to 80 lb, is the largest living mustelid. A characteristic animal of the seacoasts, it seldom strays more than about 500 yards either onto land or out to the open sea. Its long, compact coat helps to maintain constant body temperature.

Sea otter
(*Enhydra lutris*)

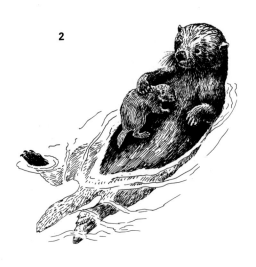

The sea otter's favourite position is floating on its back. This it may do either to sleep (1), when it wraps a frond of kelp around itself as an anchor, or when transporting its baby for short distances (2).

sharks. Man, of course, remains a constant threat, despite prohibitions on the sale of otter fur, and is responsible for a high death toll. Fishermen who accuse them of feeding so copiously on crustaceans as to jeopardise their livelihood have no qualms about killing them, regardless of the law. Finally there is the menace to the future of the species posed by the pollution of coastal waters, a slow process of poisoning which wears away the protective cover of fur.

The world of the pinnipeds

The most familiar mammals of the seacoasts are undoubtedly the members of the suborder Pinnipedia, animals whose feet have been converted into flippers (hence 'wing-' or 'fin-footed'). They spend the greater part of their life in water but reproduce on land. Zoologists divide them into three families, based on differences of form and structure—Phocidae (true seals), Odobenidae (walruses) and Otariidae (sea-lions and fur seals).

The sea-lions and fur seals are also known as eared seals because of their visible external ears (in contrast to other families in which the ear pinna is very small and hidden within the ear opening). Another characteristic is that the hind limbs, can be directed forwards like those of land mammals. This helps them to climb rocks and move rapidly over the ground. They are excellent swimmers too, stretching the hind flippers behind them, like a tail, and propelling themselves swiftly through the water with the front flippers.

The walrus resembles the Otariidae in that its hind flippers are similar in structure but the ears are hardly in evidence. Its most obvious features are the huge ivory tusks, in both sexes. It is apparent from its enormous size that it is not a very agile animal but it is far more in its element in the sea than on land.

As for the so-called true seals, their hind flippers face permanently towards the rear, like the tail, and are of no use for walking. On land they move along with little wriggling leaps, supporting themselves on the belly. Although somewhat clumsy, this method of locomotion can be quite speedy at times. But it is in the water that they have no rivals. Instead of resorting to the forelimbs for swimming, they propel themselves with fish-like movements of the hind flippers. This enables them to cut through the water very rapidly and to make sudden, sharp changes of direction.

It was originally thought that all pinnipeds stemmed from the same ancestral stock, but this theory is nowadays seriously challenged. It is possible that the apparent similarity between seals, walruses and sea-lions is simply another example of convergent evolution. The varying methods of locomotion—whether on land or in the water—are enough to suggest that they do not constitute a homogeneous group. Some modern authors claim that sea-lions, and possibly walruses too, are related to bears, and that the seals may either have a common ancestor with the mustelids or even be descended from the latter group, as a form of highly perfected sea otter.

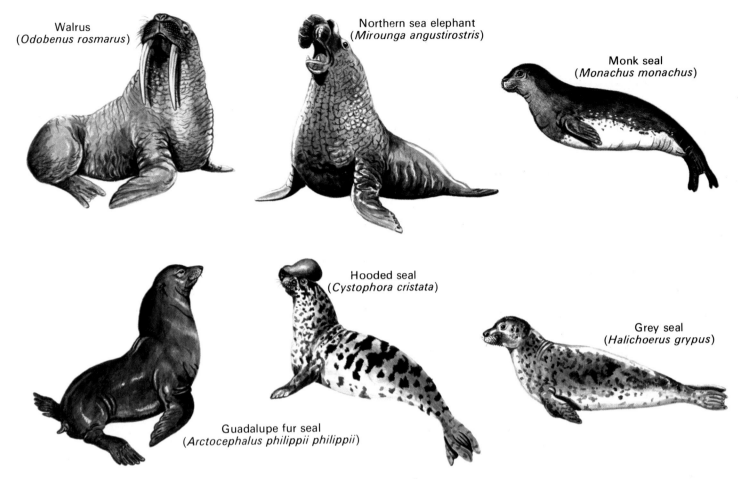

Problems of sea-diving mammals

The pinnipeds, particularly the true seals (for it is they which have been studied in greatest detail) are capable of spending lengthy periods under water, but as a rule they vary their daytime activities, bouts of swimming and fishing alternating with intervals of rest on shore. It is quite obvious, therefore, that they are ideally suited to an amphibious existence but that this has necessitated the acquisition of special physiological mechanisms for leading such a double life.

The chief problems confronting these mammals are to procure oxygen and fresh water, and to ensure that the body temperature stays constant whether they are submerged in the freezing ocean or basking on rocks in the sun. Additionally, when fishing at great depths, as often happens, they face the same difficulties as those experienced by divers.

A pinniped has no difficulty in breathing above the water surface and all of them do this quite normally when they are awake. But how do they manage it while asleep on the high seas? Principally thanks to a layer of fat so thick that it serves as a kind of inner tube, some seals are capable of sleeping in a vertical position, rather like a floating bottle. In this way their nostrils protrude above water and they can breathe regularly. Others, however, sleep in shallow water and rise to the surface from time to time, without waking, to take several deep gulps of air before sinking down again. It has even been confirmed that certain species breathe like this, in fits and starts, while

Although they spend most of the year in the open sea, seals are generally regarded as typical inhabitants of the seacoasts for most venture onto dry land to give birth, rear their young and moult.

sleeping on land. The northern sea elephant, for example, is in the habit of suspending its breathing for about eight minutes; then, for the next five minutes or so, it inhales and exhales some thirty times.

A supply of fresh water is necessary for the metabolic processes of marine mammals, and scientists have not so far discovered exactly how this is obtained. It is clear that many pinnipeds do not have regular access to fresh water and that drinking sea water may be dangerous because of the high quantities of dissolved salts entering the bloodstream. It appears that seals obtain enough water from the fish they eat. They may sometimes take in salt water with their food but in such small amounts that their bodies can cope with the extra salt.

Many pinnipeds live in areas where the water is very cold, often icy, for parts of the year, yet succeed in keeping their body temperature within acceptable limits. This is possible because the small arteries conducting blood to the skin automatically close when the animal is exposed to severe cold, and that thanks to the fatty tissue insulating the body, there will be no great internal fluctuations even when the skin temperature is not much more than 0°C. But if the animal needs to lose body heat by irradiation, as when it is basking on shore in the sun, the arterioles dilate and the blood flows freely to the surface of the body, with a consequent cooling effect. Some sea-lions are covered with long thick fur and this too helps to regulate body temperature.

Champions of deep-sea fishing

Enthusiasts of underwater fishing, an activity in which the pinnipeds undoubtedly take the prize, are well aware of the problems involved in diving. Basically there are three—an adequate supply of oxygen, effective resistance to increasing water pressure and, in the case of dives to considerable depths, returning safely to the surface without succumbing to what is commonly known as the 'bends'—occlusion of blood vessels caused by bubbles of nitrogen. Divers avert this hazard by coming up to the surface by slow stages. Yet Walker has cited the instance of a Weddell seal which dived to a depth of approximately 2,000 feet and another that stayed submerged for a period of forty-three minutes and twenty-five seconds, apparently rising to the surface without varying its normal swimming pattern. Although few seals can match such performances, all of them seem to dive and ascend again quite effortlessly. How do they accomplish this? The mystery has been partially clarified by scientists who have studied the behaviour of common and grey seals, particularly the latter, both species having been subjected artificially to strong pressures, comparable to those they would normally have to contend with in the ocean depths.

Seals do not fill their lungs with air prior to diving, but empty them almost entirely, this being much easier than in the case of other mammals by reason of the position of diaphragm and ribs. In this manner the volume of the lung cavities—those parts of the body most vulnerable to high pressure, since the rest is practically incompressible—is reduced. Furthermore, by ridding the lungs of air (composed as this is of oxygen and nitrogen) there is little risk of the latter gas being responsible for fatal embolisms.

The cause of such accidents is roughly as follows. When a mammal is subjected to crushing pressures, not only oxygen but also nitrogen is dissolved in the blood. The former is consumed by the tissues, the latter remains in the bloodstream. Then, when the pressure suddenly decreases, the nitrogen reverts to its gaseous state and forms bubbles which, by obstructing the circulatory system, provokes dangerous stoppages. That is why decompression has to be a gradual process, with the diver coming up very slowly, with halts, to the surface. It appears clearly, in the light of these explanations, that because a seal eliminates air from its lungs there can be no risk of it succumbing to the 'bends'. But how can it possibly dispense with oxygen?

When a seal dives its cardiac activity is reduced to one-tenth, even one-fifteenth, of normal, and the circulation is slowed down to a minimum. But the brain and the heart continue

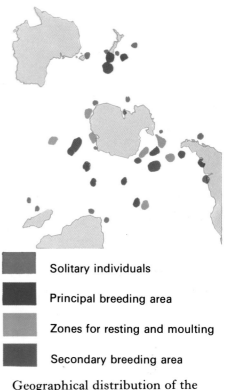

Solitary individuals

Principal breeding area

Zones for resting and moulting

Secondary breeding area

Geographical distribution of the elephant seal.

SOUTHERN ELEPHANT SEAL
(*Mirounga leonina*)

Class: Mammalia
Order: Pinnipedia
Family: Phocidae
Total length: male 175–256 inches
(450–650 cm)
female 120–138 inches
(300–350 cm)
Weight: male up to 7,700 lb (3,500 kg)
female up to 1,980 lb (900 kg)
Diet: fishes, crustaceans, molluscs, birds
Gestation: 270–350 days
Number of young: one

Largest of all pinnipeds. Elongated muzzle in form of trunk measures up to 15 inches in adult. Colour uniformly greyish-blue, paler on underparts. Baby dark brown.

Following pages: Walruses use their tusks principally for scooping molluscs and other forms of food from the ocean bed; but the tusks also pierce snow and ice to support the animals as they drag themselves across the frozen wastes.

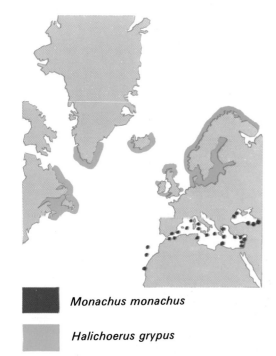

Geographical distribution of monk seal (*Monachus monachus*) and grey seal (*Halichoerus grypus*).

TRUE SEALS

Class: Mammalia
Order: Pinnipedia
Family: Phocidae
Diet: fishes, some crustaceans, molluscs, birds
Gestation: 270–350 days
Number of young: usually one

GREY SEAL
(*Halichoerus grypus*)

Total length: 71–130 inches (180–330 cm)
Weight: 275–615 lb (125–280 kg)

Colour very variable, usually darker on back than on underparts. Large pointed muzzle, especially in male, with white markings on dark ground; on muzzle of female the marks are darker and the background colour paler. Newborn baby white.

MONK SEAL
(*Monachus monachus*)

Total length: 90–150 inches (230–380 cm)
Weight: 660–705 lb (300–320 kg)

Back dark grey or chestnut. Prominent white mark of variable size on underparts. Newborn baby black.

---

to receive supplies of oxygenated blood stored in the inferior vena cava, whereas the other viscera and the limbs are temporarily starved of blood, as a result of the closure of the large and small arteries. There is no question of a risk of the blood coagulating after such a period of stagnation, and fishermen who have killed seals confirm that this only occurs later, suggesting a normal blood flow.

That, however, is not the whole story. When the mammal dives or swims under water it obviously expends energy, normally produced by the combustion of sugars with oxygen to produce carbon dioxide and water. But when oxygen is absent this energy has to be supplied in another way. It is now known that the breaking down of these sugars or carbohydrates proceeds in two stages, both of which supply energy. In the first, without any need of oxygen, lactic acid is formed; in the second, this lactic acid, together with oxygen, releases carbon dioxide and water vapour. When the seal is submerged only the first part of the reaction comes into play and the oxygen supply is assured when the animal comes to the surface to breathe. The same process occurs among land mammals and in man when the organism is subjected to violent exercise. Breathlessness after intense activity is simply due to shortage of oxygen and resultant aches and pains to the accumulation of lactic acid in the muscles. Seals tire less quickly because they are capable of tolerating large amounts of lactic acid and carbon dioxide.

Not only are seals magnificent underwater swimmers, but they are also highly accomplished hunters. The limbs are converted into flippers, the body is streamlined and all projections have either been smoothed out or—like the teats of the females and the scrotum of the males—are concealed beneath layers of fat. Being such rapid swimmers, they can catch the most lively fishes. But precisely how they locate their prey is another problem not fully understood by zoologists. Dr Hobson is of the opinion that seals have good vision, even in darkness, for the eyes are pretty well adapted to light and shade. Also, since the eyes are well up on the head looking upwards, they can see the fishes silhouettes etched against the surface. The same author has proved that seals under water can likewise spot their victims by starlight, outlined against the sky. Both by day and by night they can attack and kill their prey from below.

This may be the hunting technique most commonly employed but it has been demonstrated that even blind seals are long-lived and must obviously be able to hunt successfully. The sensitive hairs or vibrissae of the muzzle probably help the animals to detect prey but they would appear to be of value only in relatively enclosed areas, as near the seacoasts, and of limited use in the open sea. Dr T. C. Poulter, however, has suggested that certain seals and sea-lions emit ultrasonic impulses and consequently locate prey by means of echo-location, like dolphins and bats. It is worth pointing out, though, that the temporal lobes of these pinnipeds (responsible for hearing) are far less developed than in dolphins. As for the sense of smell, although this is very keen, enabling mothers to recognise their young, it is useless under water, since the nostrils are then closed.

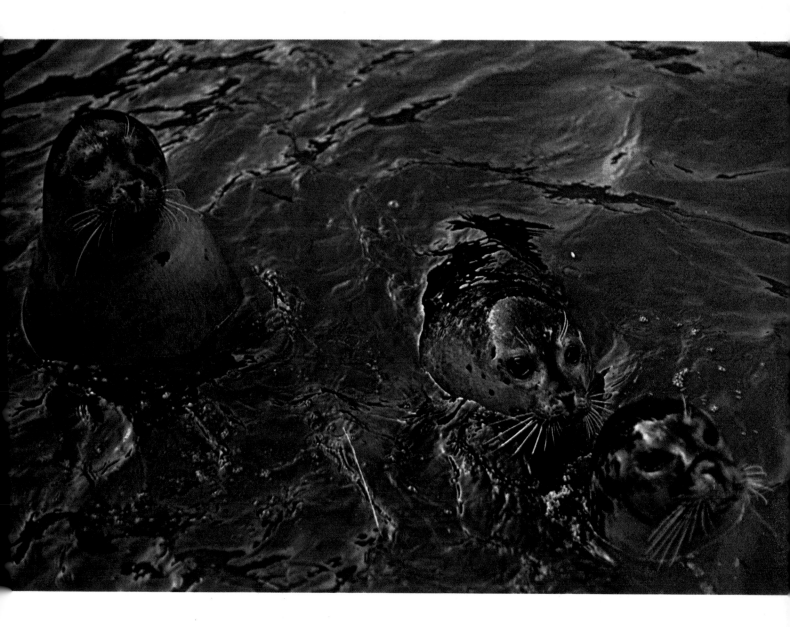

The majority of seals and sea-lions feed on fish but some base their diet on invertebrates. The leopard seal evidently kills warm-blooded vertebrates such as penguins and sea birds, but its main food is fishes and invertebrates.

Social life of the grey seal

With rare exceptions, seals leave the sea when the time comes to mate, give birth and moult. During these periods they do not eat and thus lose a considerable amount of weight. Therefore they spend as little time as necessary on dry land. Even if they were able to conduct all these activities simultaneously, entailing only a single shore visit, a great deal of energy would be expended; so among most seals there are two main periods ashore, one coinciding with the birth of the pups and adult sexual activity, the other, some months later, when shedding and renewing their coat. At both times they form compact groups that are sometimes very numerous. Each species has its own form of social organisation and within the groups themselves there are individual differences. K. M. Backhouse has made a special study of the

Seals can remain immersed in cold water for lengthy periods thanks to thick layers of fat which insulate them from their surroundings. The body temperature is thus kept within supportable limits.

Because it is an inhabitant of the coasts of northern and central Europe the grey seal has been studied in some detail. In the early days of September males and females reach the shores of the Hebrides. After the females give birth the animals mate and return to the sea until the following year. Except for the couple of weeks after the pups are born females are almost always pregnant.

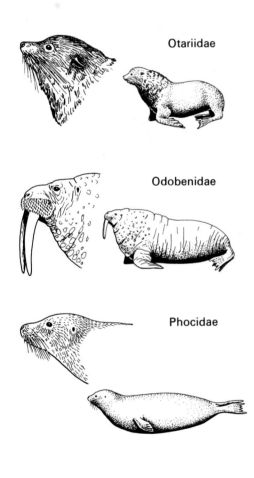

There are three clearly differentiated families of pinnipeds. The Otariidae have visible ears and hind limbs which can be directed forwards like those of land mammals. The Odobenidae can walk in a similar manner but lack visible external ears. The Phocidae also lack visible ears and their hind flippers can only be held rigidly behind them. Of the three families the Phocidae are the best adapted to aquatic life.

communal behaviour of a colony of grey seals in the Hebrides, and this survey will serve here as the basis of the behaviour of all species throughout the year.

At the beginning of September groups of males and females arrive separately on the beaches where they will mate and the females give birth. There they remain for a couple of days without mingling. Then the largest and strongest bulls wander off to take possession of the rocks between the beach and the ocean. For their part, the already gravid females begin giving birth to their pups.

Each territorial male spends most of his time basking in the centre of his domain, apparently indolent but actually on the alert. Should another male approach, his attitude will change abruptly. He stretches, flops over towards the intruder and then, with head erect and mouth wide open, lets out a series of loud roars. If, despite this warning, the stranger is stubborn and oversteps the territorial boundaries, a fight may ensue. Normally this will not produce injuries on either side, but occasionally the battle will be violent enough to cause severe wounds.

Shortly after their pups are born the females abandon them on the beach and plunge back into the sea, swimming close to the bulls' territories. Initially they return up to five times a day to suckle their offspring, and whilst attending to these maternal duties show some hostility towards the courting males, sometimes dealing out terrible bites. But after about a fortnight the visits to the young are spaced out at longer intervals and the approaches of the bulls received more amicably.

The pup of the grey seal, born with a white coat that is later shed, is fed on its mother's rich milk and puts on a couple of pounds a day until it is eighteen days old. Left alone almost as soon as it is born, except for periodic feeding sessions when the mother temporarily leaves the sea, the baby has to fend for itself very early for the female is already on the lookout for a mate.

For the most part grey seals are polygamous and the bulls rule over jealously guarded harems; but the observations of Backhouse in the Hebrides show, on the contrary, that there is also a high degree of promiscuity. The female is sexually receptive two weeks after giving birth. One day, whilst following her usual route down to the sea, she will simply choose a partner and enter his domain. Coupling takes place in the water, preceded by no special courtship ritual, and lasts between fifteen and twenty minutes. In the course of sexual activity both animals disappear several times under water, coming up at intervals to breathe. Soon afterwards the female will stray off to find a new mate.

Gestation lasts 350 days and then males and females will return to the same sites and mate once more. Thus an adult female is pregnant virtually all her life, except for two weeks every year. This synchronisation of births and mating ensures that as short a time as possible is spent on land.

Precocious pups

The newborn grey seals are already covered with a long silky white coat and several hours later are left alone on the beach. Their mothers only return periodically to feed them, each guided by her pup's scent. In crowded breeding colonies fights often break out among females to protect their endangered pups and angry bites may be inflicted on any stranger or orphan approaching an adult female as she clambers up the beach.

The pups have already lost their milk teeth whilst in the

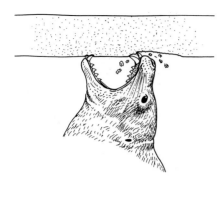

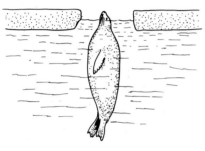

Some seals—such as the Weddell seal—do not migrate to warmer zones in the winter In order to breathe the animals use their powerful teeth to chew holes in the ice.

womb and their adult dentition shows signs of appearing a few days after birth.

As soon as she recognises her baby the mother stretches out and offers it her teats. The female seal never licks her baby, in the way land mammals do. Since the teats are hidden under the fat the pup has to make a few experimental licks not only to locate them but to stimulate their erection in order to suckle. Seal's milk is very thick and nourishing, containing 50 per cent fat, five times as strong as the milk of cattle or humans. On such a rich diet the baby grows rapidly and in little more than two weeks has trebled its birth weight of around 20 lb. In other words it puts on roughly 2 lb every day. In a sense, what it gains is lost by the mother, who refrains from eating at this time. At around eighteen days the youngster begins to acquire the characteristic coat of the adults. Soon afterwards, when this transformation is complete, the young seal, finally abandoned by its mother, will make its way down to the ocean and from then on will have to fend for itself.

Certain species are even more precocious than the grey seals, losing their first coat while still in the womb and being born with their definitive fur. This is the case with the young common seal, for example, which only a few hours after birth is already following in the tracks of its mother down to the sea. It is ready to swim soon after birth.

After abandoning the breeding colonies (the females leaving several days before the males) adult grey seals spend some months in the sea feeding voraciously in order to restore their energy. Then they return for a short period to land and undergo another fasting session while they shed their fur. After moulting the seals return to the sea until the following breeding season.

The cows reach sexual maturity at the age of three years, and the bulls at more or less the same time. If twins are born, which happens rarely, the mother usually feeds one and leaves the other to perish of starvation.

Only the elephant seal among the true seals appears to stake out nuptial territory, and the bulls are polygamous. Other seals are monogamous and pairs remain together throughout the mating season and often for the entire year.

The far-ranging seals

There are thirteen genera and eighteen species of seals which are distributed not only, as is commonly supposed, in cold and temperate waters but are also present in tropical seas and inland lakes, although there they are less abundant.

Four species of Phocidae live in the southern hemisphere. The crabeater seal (*Lobodon carcinophagus*), belying its name by consuming krill and fishes, is the most numerous, yet very little known. The Weddell seal (*Leptonychotes weddelli*) has the most southerly range and remains among Antarctic ice even in winter, chewing holes in the frozen surface with its teeth. The Ross seal (*Ommatophoca rossi*) is seldom seen and not much is known about it. The leopard seal (*Hydrurga leptonyx*), a large animal

with solitary habits, certainly kills penguins, sea birds and young seals, although its main food is now known to be fishes and invertebrates.

Largest members of the family are the southern sea elephant or elephant seal (*Mirounga leonina*) and the northern sea elephant (*Mirounga angustirostris*). The northern species, almost exterminated at one time, is now protected. The two species are similar, the bulls being enormous, the cows much smaller.

Apart from the elephant seal there are nine species of seals in the northern hemisphere. The Greenland seal (*Pagophilus groenlandicus*) is the most accomplished swimmer in the family and also moves about easily on land. A true inhabitant of the Arctic is the hooded seal (*Cystophora cristata*), so named because of the swollen skin between the muzzle and eyes of the males. This species migrates long distances to escape the winter ice, unlike the bearded seal (*Erignathus barbatus*) which remains under the ice and breaks holes in the surface, through which it breathes. The banded seal (*Histriophoca fasciata*) lives in the North Pacific around the Bering Sea and is identified by the white stripes on neck and flanks, contrasting with its overall dark brown colour. The ringed seal (*Pusa hispida*) is a small species with a circumpolar range, with close relatives in the Caspian Sea (*Pusa caspica*) and Lake Baikal (*Pusa sibirica*), the Baikal seal being the smallest of all Phocidae. The common seal (*Phoca vitulina*) is widely distributed along the Pacific and Atlantic coasts while the grey seal (*Halichoerus grypus*), a familiar inhabitant of northern parts of the British Isles, is a North Atlantic species.

Finally there are three seals of the genus *Monachus*, all residents of warm seas. The monk seal (*Monachus monachus*) of the Mediterranean, a victim of man's hunting, is most often seen today in the Black Sea and along the coasts of West Africa as far as Mauretania. The Caribbean monk seal (*Monachus tropicalis*),

Still abundant, though little known, crabeater seals (which eat only fishes and krill) live on the Antarctic ice-cap. It is thought to be a semi-gregarious, nocturnal species, reaching sexual maturity between the ages of three and six years.

Although capable of swimming soon after they are born, baby walruses remain with their mothers for one and a half to two years. Females therefore reproduce only once every two or three years.

The tusks of the male walrus (*left*) are larger and heavier than those of the female. They are used for fighting (1), for walking over ice (2) and for extracting food from the sea bed (3).

abundant in the 19th century, is in danger of becoming extinct, but the Hawaiian monk seal (*Monachus schauinslandi*), once on the verge of disappearing, seems to have made a recovery, its population now numbering about 2,000 individuals.

The inoffensive walrus

The walrus, a massive, barrel-shaped marine mammal with greyish-brown fur and long vertical ivory tusks, is an inhabitant of Arctic waters. Zoologists distinguish two subspecies, the Pacific walrus (*Odobenus rosmarus divergens*) and the Atlantic walrus (*Odobenus rosmarus rosmarus*). In both races the bulls, larger and stronger than the cows, may weigh up to 3,500 lb in winter when the bulk of the body is increased by a thick layer of fat. The Pacific walrus is also heavier than the Atlantic subspecies.

The tusks are enormous. In males of the Pacific race they may measure more than 3 feet, those of the females being up to 2 feet long. These tusks, which are continuously growing upper canines, are more markedly curved in the females. They are not, of course, merely ornamental and may serve, should the need arise, as redoubtable weapons. But these huge mammals, which live in large herds numbering about a hundred individuals, are basically timid, inoffensive and fairly peaceful creatures. Only if directly menaced are they liable to turn on their attackers and there have been recorded cases of their overturning small boats used by hunters.

The tusks also have an important function in helping the walrus to walk over ice, the sharp tips supporting the animal much as crampons do a mountaineer. This attribute, incidentally, accounts for the generic name, for *Odobenus* literally means 'one who walks with his teeth'.

The main value of the tusks, however, is to probe for food, this consisting principally of molluscs, sea urchins and crustaceans, these small animals being scooped efficiently from the mud of the ocean bed. The long, conspicuous bristles on the muzzle, with roots richly furnished with blood vessels and nerves, are also extremely sensitive and undoubtedly play an important part in the location of such prey. As a rule the walrus procures its food in comparatively shallow water and does not dive much deeper than 200 feet; but occasionally it may descend to 300 feet. Scientists are not certain how the animal manages to withstand the water pressure at such depths but conclude that the physiological mechanisms that come into operation are similar to those possessed by seals (see pages 109 and 112).

Some individuals, generally older males, tend to modify their normal food habits and become carnivorous. They begin by eating carrion and having, as it were, acquired the taste for flesh, may even kill young seals.

At five or six years of age the males reach sexual maturity, the females having done so about a year previously. Mating occurs from April until the end of May and gestation lasts a few days less than a year. The single pup, covered with short grey fur and weighing about 100 lb, is born on the ice. Its mother quickly leads it to the water and teaches it to swim, an activity which the baby does not properly master until it is a couple of weeks old. The mother jealously protects her rapidly growing youngster and does not allow it to wander from her side for at least a year and a half, often two years. During this entire period the mother continues to suckle her offspring and even after it is weaned will not leave it alone for several months more. This explains why females only give birth once every two or three years.

As is the case with seals, walruses are of enormous importance to the Eskimos, who find a use for almost every part of the body—meat, fat, fur, teeth and bones. But it has been the precious ivory tusks which have led to the intensive hunting that has almost brought about the extinction of the Atlantic walrus. Although the export of tusks and fur is now prohibited and protective measures have been enforced by many governments, more than 2,500 animals are still killed every year by Eskimos, now equipped with rifles and motor boats, and this toll is too high in relation to the low birth rate and known population counts. Apart from man, the worst enemies of walruses are killer whales, whose appearance among a herd is enough to provoke immediate, panic-stricken flight.

The eared seals

The eared seals of the family Otariidae fall into two groups, distinguished mainly by the quality of their coat. The sealions (Otariinae) have short, coarse hair which is of little commercial value; but the fur seals (Arctocephalinae) are covered with thick, soft hair which has always been highly prized

Odobenus rosmarus rosmarus
Odobenus rosmarus divergens

Geographical distribution of Atlantic walrus (*Odobenus rosmarus rosmarus*) and Pacific walrus (*Odobenus rosmarus divergens*).

WALRUS
(*Odobenus rosmarus*)

Class: Mammalia
Order: Pinnipedia
Family: Odobenidae
Total length: 118–178 inches (300–450 cm)
Weight: 1,760–3,500 lb (800–1,600 kg)
Diet: molluscs, crustaceans, fishes
Gestation: 360 days
Number of young: one

Female smaller than male. Colour light brown. No visible tail. Long vibrissae of muzzle help to locate food. Long, powerful tusks in both sexes, sometimes up to 3 feet long in male, 2 feet in female. Hind flippers can be directed forwards to assist walking. Pups greyish, weighing about 100 lb at birth, are fed for one and a half to two years by mother, until tusks are large enough. Two subspecies are Pacific walrus (*Odobenus rosmarus divergens*) and Atlantic walrus (*O. r. rosmarus*).

CALIFORNIAN SEA-LION
(*Zalophus californianus*)

Class: Mammalia
Order: Pinnipedia
Family: Otariidae
Total length: male 93 inches (235 cm)
female 71 inches (180 cm)
Weight: male 615 lb (280 kg)
female 198 lb (90 kg)
Diet: fishes, octopuses, squids
Gestation: 340–365 days
Number of young: one

Brown coat, tending to appear black when wet. Top of head becomes slightly paler with age. Prominent ears. Hind flippers capable of turning forwards for walking. Newborn pup able to swim very early and is suckled for a year.

by the fur trade. Of the two subfamilies it is the sea-lions which are better known. One of the species, the Californian sea-lion (*Zalophus californianus*), wrongly and misleadingly described as a 'seal', is the familiar animal of circuses and zoos which balances rubber balls on its nose and can be taught to perform other spectacular and entertaining tricks.

Sea-lions live in enormous mixed herds on rocky seacoasts and some species are migratory. These groups disperse during the breeding season when the powerful males lay claim to territories and harems which they guard against all rivals. As happens with true seals, birth coincides, more or less, with mating. The females, shortly after giving birth, return to the sea, only coming back to shore two or three times daily to suckle their young.

When sea-lions put in an appearance near fishing grounds they are usually hunted without mercy for they feed on all kinds of fishes, including salmon when they return to river estuaries to spawn. But the Californian sea-lions seem to prefer squids, cuttlefish and octopuses. Steller's sea-lion (*Eumetopias jubatus*), an inhabitant of the Pacific and Arctic coasts of Asia and North America, eats vast quantities of lampreys, and since the latter are a scourge of salmon such food habits can be said to be of benefit to the fisheries.

Other species are the Australian sea-lion (*Neophoca cinerea*), the New Zealand sea-lion (*Phocarctos hookeri*) and the South American sea-lion (*Otaria byronia*).

The fur seals are typical inhabitants of the southern hemisphere where six of the seven surviving species live, all belonging to the genus *Arctocephalus*. The other is the Alaska fur seal or sea-bear (*Callorhinus ursinus*) of the North Pacific. Diet and reproductive habits are similar to those of sea-lions.

Geographical distribution of different species of sea-lions.

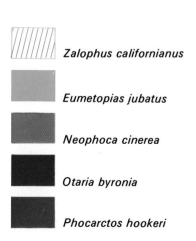

Zalophus californianus

Eumetopias jubatus

Neophoca cinerea

Otaria byronia

Phocarctos hookeri

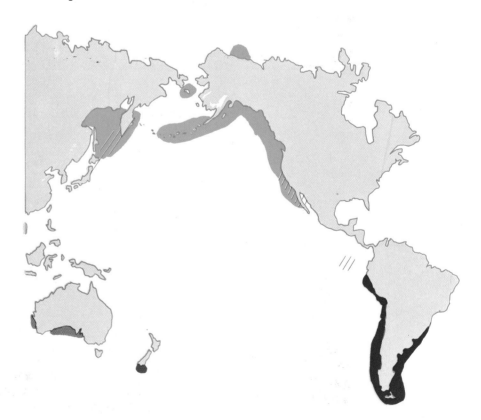

It is worth noting that a previously unknown species of deep-sea fish, *Bathylagus callorhinus*, was found for the first time in the stomach of a sea-bear.

Hunted pitilessly for their pelts, fur seals have been slaughtered in their hundreds of thousands. The Guadalupe fur seal (*Arctocephalus philippii philippii*), once found in large numbers among the islands of the Juan Fernández archipelago, off the coasts of central Chile, has been only one tragic victim of this inhuman trade. When the explorer and buccaneer, William Dampier, visited the islands in the 17th century he described how virtually every bay and rock teemed with these plump seals, and spoke of the opportunities awaiting those sea-going nations capable of exploiting them for fur and oil. His prediction was accurate. By 1901 the species had all but disappeared.

In November 1968, however, Kenneth S. Norris and William N. McFarland found about thirty of these seals (of both sexes and varying ages) in the rocky caves of the island of Más a Tierra. According to local fishermen there are more animals on the nearby island of Más Afuera. The Chilean government has taken steps to protect these seals and although numbers are modest there are hopes for their survival.

The imperilled monk seals

The Atlantic walrus and the Guadalupe fur seal are unfortunately not alone in their plight. At least five other pinnipeds, hunted for their flesh, fur and fat, are also listed in the IUCN Red Data Book of mammals in danger of extinction. Among them

Sea-lions live in large colonies which break up for the breeding season. Each bull then marks out a territory, defending it against all rivals and attracting a harem of cows.

Fur seals (*above and below*) have been remorselessly hunted for their short, soft fur, so much so that their populations have been alarmingly reduced; however, attempts are now being made to reintroduce them to islands that they once inhabited.

are the three monk seals of the genus *Monachus*, including the Mediterranean species.

According to information provided by Wijngaarden, referred to by Noel Simon, there would appear to be a small but fairly stable community of monk seals off the shores of Greece and adjacent islands. One solitary animal has been sighted close to the Yugoslavian coast and there is another tiny, isolated group in a cave in north-western Corsica. In Spain there are thought to be a few surviving monk seals between Alicante and Almeria; and in 1960 a dead seal was found stranded on the northern coast of Majorca. In Roumania there are a few individuals left in the delta of the Danube, and two small colonies still survive on the coast of Bulgaria.

The balance sheet appears to be a little more favourable on the coasts of Turkey, Cyprus and Lebanon; and one of the larger colonies is to be found among the caves of the Bay of Alhucemas in Morocco. Outside the Mediterranean there may be small groups on the Canary Islands and on the shores of Madeira; but the most important community flourishes off the south-western tip of Spanish Sahara.

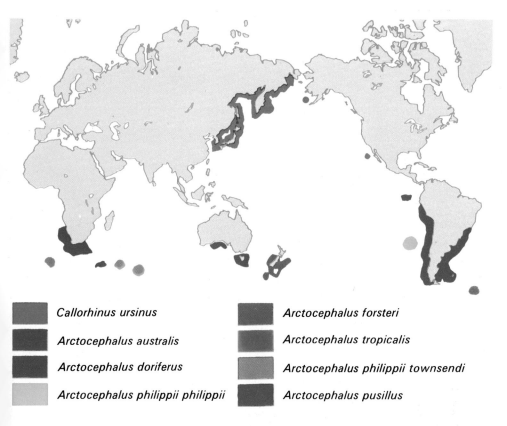

Geographical distribution of different species of eared seals.

- Callorhinus ursinus
- Arctocephalus australis
- Arctocephalus doriferus
- Arctocephalus philippii philippii
- Arctocephalus forsteri
- Arctocephalus tropicalis
- Arctocephalus philippii townsendi
- Arctocephalus pusillus

ALASKA FUR SEAL
(*Callorhinus ursinus*)

Class: Mammalia
Order: Carnivora
Family: Otariidae
Total length: male 75–83 inches (190–210 cm)
female 59–67 inches (150–170 cm)
Weight: male 396–660 lb (180–300 kg)
female 77–150 lb (35–68 kg)
Diet: fishes and some molluscs
Gestation: about one year
Number of young: one

Very compact fur, male being dark grey on back, reddish-brown on limbs and belly, and female usually greyer. At birth the blackish pup weighs about 10 lb.

In all there are probably only about 500 Mediterranean monk seals left alive, and most of these, in order to escape the fate of countless others of their species, have modified their normal behaviour patterns. Instead of basking on open beaches they now live only in caves with underwater entrances. The future prospects are not very hopeful, although a promising start could be made if the Alhucemas and Spanish Sahara populations, at least, were afforded official protection.

The extinct sea-cow

The marine mammals of the order Sirenia, the manatees and dugongs, are inhabitants of the coastal waters of tropical seas, often finding their way into estuaries and far up rivers. Steller's sea-cow (*Hydrodamalis stelleri*) was the only representative of the order ever to have lived in colder climes, for this animal, largest of all sirenians, frequented the Bering Sea. It was discovered in 1741 yet within about twenty-seven years it had to all intents completely vanished.

Indications are that this animal was very large, probably about 25 feet in length and weighing something like 4 tons. All the information concerning its biology was obtained from the writings of the German naturalist Steller, after whom the animal was named. Apparently it led a tranquil life in family groups that consisted of a male, a female and two pups. Certainly the species must have been monogamous, the babies being born at any time of year but generally in the autumn. Food consisted mainly of marine phanerogams.

The slow movements and passive nature of Steller's sea-cow made it an easy prey; and since its flesh was edible and its fur of value, it was probably hunted to extinction.

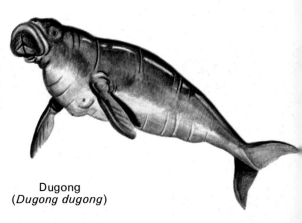

Dugong
(*Dugong dugong*)

CHAPTER 6

The gorgeous world of the coral reef

For centuries poets have extolled the splendours of the ocean yet their ecstatic descriptions have been based on a very tiny known area of the sea's surface. By far the most beautiful and spectacular regions of the ocean lie hidden below the waves, though not at great depths, and it has only been quite recently (since the end of the Second World War) that man has seriously begun to explore them. Prior to that professional divers, clad in traditional diving suits, had seldom ventured beyond the turbid waters of bays and harbours. Yet in the previous century the French writer Jules Verne, in his *Twenty Thousand Leagues Under the Sea*, had brought a powerful imagination to bear on the probable mysteries of the deep. Inaccurate in many details, his vivid descriptions are not much less fantastic than the realities of the 'silent world' as revealed by Commander Jacques Cousteau and other daring underwater explorers.

Quite apart from the more sophisticated equipment of the scientists, a miscellany of masks, snorkels, flippers and oxygen canisters have been manufactured in recent years which enable sportsmen and holiday-makers to savour some of the wonders of the underwater kingdom. Those who have had the good fortune to pursue their hobby in tropical seas—the Caribbean, the Red Sea, the Pacific or the Indian Ocean—will admit, however, that for sheer beauty nothing can compare with a coral reef. Indeed this discovery has brought peril in its wake. The coral reef of Malindi, on the coast of Kenya, for example, is now a national park, protective measures having become essential because of the threat posed by underwater fishermen to its rare species. Today the fishes of these coral reefs tolerate the presence of skin-divers armed, not with harpoon guns, but with submarine cameras. The theft of so much as a shell from

Facing page: Probably the most striking feature of the underwater world of the coral reef is the dazzling beauty of the many fishes that make it their home. This handsome individual is a blue faced angelfish *Euxiphipops xanthometapon*.

Reef-building corals cannot develop at low temperatures so they are only found in tropical seas, usually off east coasts.

Facing page: The strange formations of coral reefs, together with their magnificently coloured inhabitants, are treasure troves for scientists, with a diversity of flora and fauna unequalled in any other marine habitat.

Malindi, let alone a fish, is subject to heavy penalties, and such severity is well justified. Were hunting activities not controlled the coral reef would literally die, the sea anemones vanishing, the colourful fishes fleeing to safer, more hospitable waters.

In the translucent tropical seas where true corals grow, the most striking impression is the remarkable range of colours displayed by the fishes that dart in and out of the branching corals—blending and contrasting hues, subtle nuances and bizarre patterns that would challenge the ingenuity of any painter (not for nothing is *Rhinecanthus aculeatus* named the Picasso fish!). But as zoologists are well aware, such brilliant tones, allied in many cases to curious body shapes, have definite purposes. They are necessary not only for delimiting territory but also for intraspecific recognition. In these well-lit waters fishes can see one another clearly and immediately distinguish the visual signals which avoid confusion and which, among such dense populations, are vital to the perpetuation of the species. These markings are in a sense their war colours, serving to avoid bloodthirsty encounters among members of the same species and to warn off other species hunting food in the same zones. That such fishes have good vision is certain, but whether they can distinguish colours, as against simple outlines and contrasted markings, is not always certain.

Sometimes there are other reasons for these glorious colours. Certain species, for example, possess vivid tentacles or are covered with venomous spines. In their case the colours are clear warnings to passers-by that contact with these appendages can spell mortal peril. Furthermore some fishes have instantly recognisable patterns which are virtually 'safe-conduct passes', for they associate closely with dangerous predators. Such is the case with the pilot fishes that swim freely around sharks, for they have broad black, dark blue or violet stripes that contrast with a lighter blue background. The cleaner wrasses, ridding larger species of parasites, possess similarly distinctive striped patterns; and so do the clown or anemone fishes which swarm around the tentacles of sea anemones, immune to their poisonous stinging-cells.

The more brightly coloured coral fishes never venture too far from the relative safety of the reef, with its countless cavities and hiding places, for in the open sea they would be too conspicuous, immediately attracting the attention of carnivores. With their compact, often rounded bodies, they are experts in the game of underwater hide-and-seek, squeezing into the narrowest openings and almost impossible to catch over short distances. Those species that are less agile go to the opposite extreme. They find safety not in brilliance of colour and pattern but in plain drabness, merging so perfectly with rocks, algae or sand that even an experienced diver cannot make them out. Some poisonous species are responsible for a number of fatal accidents every year, looking so like stones that unsuspecting bathers are liable to step on them by mistake.

Finally there are a number of gregarious, non-territorial fishes that habitually swim in shoals. These are generally found

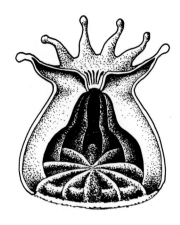

A coral reef is constructed mainly by a multitude of colonial invertebrates with chalky skeletons. The most typical of these are the coral polyps, a diagram of which is shown here. The mouth is surrounded by tentacles with stinging-cells and the tiny animal feeds on zooplankton.

The Great Barrier Reef, stretching parallel to the north-east coast of Australia for about 1,250 miles, is probably the most famous and spectacular coral formation in the world.

on the fringes of a reef and beyond in the open sea. The contrasting colours or stripes of such fishes, when seen in certain lights, probably have the effect of breaking up their silhouette (as in the case of zebras on the savannah), thus helping to deceive the enemy.

Tiny architects of the reef

Although it resembles a forest of many-coloured flowering plants, the coral reef is in fact largely of animal origin. The component parts of these colossal, fantastically shaped structures are the calcareous skeletons of millions of tiny, primitive coelenterates, especially coral polyps.

The term 'coral' is loosely used to describe a variety of organisms that differ greatly from one another in shape, colour and growth. But basically they are all cnidarians (thus belonging to the same group as jellyfishes and sea anemones), with a solid, supporting chalky skeleton. Some, particularly those of temperate seas, are solitary, living in deeper water and lacking reef-building properties; but others, chiefly in tropical seas, form immense colonies. It is the latter which make up the familiar reefs that are exposed at low tides.

The individual polyp is seated upon a tube-shaped skeleton with an unmovable supporting base. This is built up by cells that extract calcium carbonate from the sea and transform it into limestone. The polyp is basically two-layered, the body wall being composed of two walls of epithelial tissue, the inner of which is a digestive tissue. Between the inner and outer layers of epithelium is a mesoglea from which the sex cells arise. Other special cells on the tentacles surrounding the mouth of the animal contain a toxic substance that may be painful, sometimes dangerous to man. It is by means of these stinging-cells that the coral procures food.

Corals are for the most part hermaphrodite, with fertilisation occurring within the stomach. A ciliated larva develops from each egg and after leading for a while a planktonic existence eventually attaches itself to a rock or other solid surface. In colonial corals growth is effected by means of successive buddings when the polyps mature.

Special conditions have to be present in order for a coral reef to develop. The water temperature must be at least 20°C (68°F). There must be no sediment and plenty of light. The water must be in constant movement and the salinity equivalent to that of the ocean, for a diminution of more than twenty-five parts per thousand is not tolerable. Because of these exigencies coral reefs can only grow in tropical latitudes and on the eastern shores of continents and islands. On west coasts cold winds prevent their development.

Since light is absolutely vital, reef corals live in shallow water only and seldom deeper than about 150 feet. Furthermore, it would appear that for reef-building the colony must enter into symbiotic association with certain one-celled algae known as zooxanthellae. The function of these plants is not completely understood. It was once thought that the algae

might supply oxygen, but this theory has now been discarded since there are obviously easier ways for the coral polyps to procure this. What seems probable is that the algae absorb the waste matter of the coral and furnish unidentified food substances which are in some ways analogous to vitamins and hormones. The degree of dependence on these algae varies according to the type of coral. Some die very quickly in darkness (in which case photosynthesis is impossible and the zooxanthellae cannot survive) whereas others can live for several weeks under such conditions. But certainly the absence of algae has a debilitating effect on the coral colony. Tests have shown that normal development is reduced to about one-tenth and that surviving corals grow very slowly and are extremely fragile. There is no foundation in the former belief that the polyps consume these plant organisms for they feed exclusively on zooplankton and do not possess the enzymes necessary for breaking down and digesting cellulose. Yet zooxanthellae have been detected in the ratio of thirty thousand individuals to every cubic millimetre of living animal tissue. The density of this population (added to which are other algae living in symbiotic association with the corals, in their chalk-like branches) indicates that the

Coral reefs, composed of a complex miscellany of living organisms, harbour a wealth of plant and animal species, both on the surface and down to depths of several hundred feet. The waters in which they develop are transparent but always in movement, essential requisites being a high temperature, a salt content equal to that of the ocean and plenty of light.

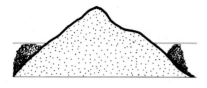
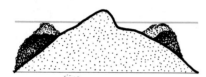
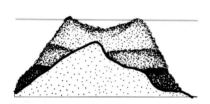

Various theories have been put forward to explain the probable origins of a coral island or atoll. According to Charles Darwin (whose hypothesis is still widely accepted) a coral reef forms around a volcanic island and continues to grow as the island gradually subsides (because of submarine upheavals and changing water levels). All that finally remains is a ring of coral around a lagoon.

Many coral fishes have acquired long snouts which enable them to probe among the fissures for invertebrates or to remove their prey from the protective cover of stinging polyps.

coral reef, contains a higher proportion of plant to animal matter than is generally supposed. These symbiotic algae may also facilitate and speed up limestone metabolism.

It is interesting to note that coral polyps are not the only constituents and builders of these huge reefs. Among a multitude of other organisms are calcareous algae which because of their tiny size fill the gaps left by corals and cement the framework into a solid mass. Furthermore, a coral reef must not be envisaged as a passive structure but as a dynamic organism in which it is possible to distinguish (as in other simple animal forms) anabolic and catabolic processes. The former are represented by the mere growth of the corals and associated organisms; the latter are due to purely physical factors (such as the destruction of portions of colonies by waves and tides) as well as to biological causes. Thus a large number of animals exercise a destructive influence on the coral reef; they include boring bivalve molluscs, burrowing worms, sponges and also fishes. There are, for example, specially adapted fishes that feed either on the individual polyps or on the entire coral, gnawing away the limestone to obtain the more nutritious soft parts. Certain sea cucumbers feed in a more selective manner, ingesting detached or dead pieces of coral, absorbing the nutritive substances and eliminating the rest in the form of a fine, pure calcareous sand. One might assume that the amounts of such waste matter would be infinitesimal, but in fact a single sea cucumber is capable of producing from 20 to 175 lb of detritus each year!

Gwynne Vevers has pointed out that a coral reef attracts and concentrates life in the same way as a clump of trees in open cultivated land or an oasis in a desert. There is no space here to do more than hint at its complex ecological pattern; but it is worth noting that the vast quantities of algae removed by plant-eating fishes help the coral to grow. And here, as in other regions, the chief threat to prosperity is man, with his pollutants and offshore industrial activities.

Coral islands

There are three principal types of coral formations. Fringing reefs jut out directly from the coast; barrier reefs run more or less parallel but some distance from the shore, from which they are separated by a navigable lagoon (the most famous example being the Great Barrier Reef, extending for some 1,250 miles off the north-east coast of Australia); and finally there are coral islands or atolls. These islands are roughly circular or horse-shoe-shaped (enclosing or partially enclosing a lagoon) and tend to appear in remote parts of the ocean, without any apparent relationship to submerged land masses. Sometimes the foundations of these immense limestone formations go down 300 feet or more, far beyond the depths at which reef-building corals can live. How then have these atolls been built?

There have been any number of theories to explain the phenomenon but the most likely one remains that advanced by Charles Darwin more than a century ago. In the course of his famous

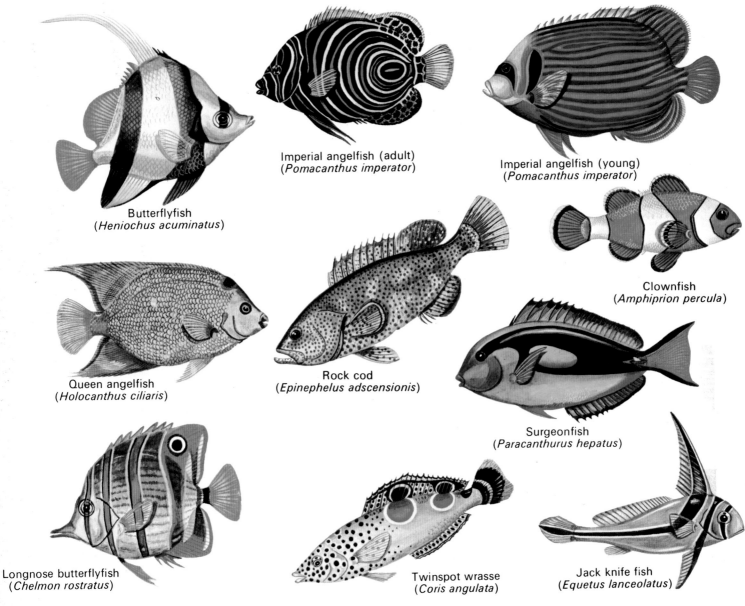

Butterflyfish (*Heniochus acuminatus*)
Imperial angelfish (adult) (*Pomacanthus imperator*)
Imperial angelfish (young) (*Pomacanthus imperator*)
Queen angelfish (*Holocanthus ciliaris*)
Rock cod (*Epinephelus adscensionis*)
Clownfish (*Amphiprion percula*)
Surgeonfish (*Paracanthurus hepatus*)
Longnose butterflyfish (*Chelmon rostratus*)
Twinspot wrasse (*Coris angulata*)
Jack knife fish (*Equetus lanceolatus*)

voyage around the world in the *Beagle*, and as a result of direct observations of coral in the Cocos Islands of the Indian Ocean and elsewhere in the Pacific, Darwin asserted that the three familiar types of coral formation simply represented successive stages in the building of a reef. According to this subsidence theory, the first stage would be the development of a fringing reef around an island which, as a result of submarine upheavals and changing sea levels, gradually sinks below the waves. The reef continues to grow upward and outward to form a barrier reef as the island slowly disappears. Finally, only a ring of coral, with an inner lagoon, is left as the island vanishes. Darwin was unable to submit proof positive of his theory, but recent investigations, using modern drilling equipment and other scientific innovations, have to some extent confirmed it. Primitive rocks have been recovered of some of these vanished islands, and examination of the base of atolls has shown them to be at least twenty million years old.

Although many problems relating to reef-building remain unresolved, science has of late come up with some of the answers.

The fishes of the coral reef are notable, above all, for their extraordinary range of colours and patterns. Bright colours serve as signals, principally for purposes of intra-specific identification. In some cases they help to break up the outline of a fish so that predators will be confused and deceived. Certain territorial species, such as the imperial angelfish, possess more brilliant colours when they are young than they do in later life.

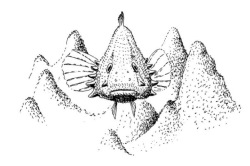

Coral fishes adapt to their surroundings in the most remarkable ways. A striking example, both in its shape and colour, is this curious individual which belongs to the genus *Caracanthus*.

The crown-of-thorns starfish (*above*) feeds on coral. When its population multiplies out of proportion considerable damage may be done to a reef. The monster mollusc of the coral reef is the giant clam (*below*), measuring 3–4 feet. Like corals, this animal lives in symbiotic association with algae known as zooxanthellae which live in the margins of the mantle.

One long-standing mystery, for example, was an explanation for the presence of a network of canals (in the majority of atolls) linking the enclosed lagoon with the open sea, thus breaking the ring of coral in many places. The larger the atoll the more numerous, evidently, were the canals. Comparative studies have now demonstrated that these breaches in the coral correspond to the mouths of freshwater rivers and streams that once criss-crossed and irrigated the primitive island. Since these now-vanished freshwater currents would have reduced the salt content of the sea in that area, corals would not have been able to develop. This process would have continued as long as a part of the island remained above the surface. Little by little the trenches sank, so that when the island was completely engulfed the beds of the canals would have subsided to a depth where no light penetrated and where no coral colonies could grow. As for the direct relationship between the number of canals and the dimensions of the atoll, this is easy to explain. The larger the island, the more rivers and valleys it must have contained, and the ring of coral and lagoon that remained would have covered the same area as the island once occupied.

Fluorescent coral

Some corals, such as those of the genus *Fungia* (so named because they resemble the convoluted underside of the head of a mushroom) live in solitude. Certain species discovered in the deep waters off New Caledonia, where only ultra-violet rays penetrate, appear to be brilliantly coloured. Yet when they are brought up to the surface these splendid hues disappear, giving way to a dull greyish tint. Intrigued by this transformation, scientists placed the corals in aquariums. When subjected to ultra-violet light they recovered their former colours, evidently as a result of a phenomenon of fluorescence. How or why they emit this fluorescent light is not known but it certainly proves that these corals, like all other species, possess pigments.

Strange invertebrates of the reef

A coral reef is the home of many wonderful animals which are not found in any other ocean habitat. Although the most obvious marvels, both from the viewpoint of unusual shape and dazzling colour, are the tropical coral fishes, many of the invertebrates of the reef are equally beautiful. Among the echinoderms, for example, are astonishing sea stars (of a kind not encountered in temperate seas), one of which, the crown-of-thorns starfish (*Acanthaster planci*), causes considerable damage to the Great Barrier Reef and to other reefs in Polynesia, because it literally feeds on coral. Hardly less colourful, but much less harmful, are primitive crinoids, otherwise known as feather stars, which grow abundantly around coral reefs. Another echinoderm the basket-star (*Astrophyton*) gradually develops a tangle of moving arms.

Molluscs are of course abundant on coral reefs. Some of them are natural works of art, especially the cowries, cone-shells

The trumpetfish has a cunning way of catching its prey, concealing itself along the back of another fish while the latter feeds and then pouncing on other fishes attracted by the feast.

TRUMPETFISH
(*Aulostomus maculatus*)

Class: Osteichthyes
Order: Gasterosteiformes
Family: Aulostomidae
Total length: up to 24 inches (60 cm)

Very long, slender, brown body, with long tube-like snout; small fins.

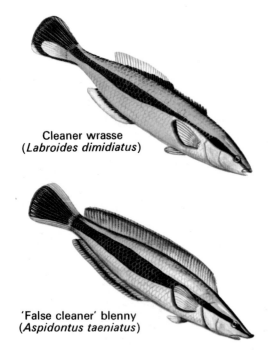

Cleaner wrasse
(*Labroides dimidiatus*)

'False cleaner' blenny
(*Aspidontus taeniatus*)

A remarkable example of mimicry is that of the 'false cleaner' blenny which has acquired almost the same colours and markings, and has adopted the characteristic movements of the cleaner wrasse, which rids other larger fishes of parasites.

CLEANER WRASSE
(*Labroides dimidiatus*)

Class: Osteichthyes
Order: Perciformes
Family: Labridae
Total length: up to 3¼ inches (8 cm)

General colour silvery blue, becoming darker towards rear. Black stripe extends along sides from snout to tail, where it broadens out. Mouth terminal.

'FALSE CLEANER' BLENNY
(*Aspidontus taeniatus*)

Class: Osteichthyes
Order: Perciformes
Family: Blennidae
Total length: about 3¼ inches (8 cm)

Slender body with coloration similar to that of cleaner wrasse. Long dorsal and anal fins, former joined to head. Mouth ventral.

and murices which are to be found in the most curious shapes, in a wide variety of delicate shades and often with extraordinary patterns.

The Indian and Pacific Oceans are the home of one amazing bivalve, the giant clam (*Tridacna gigas*), which is 3–4 feet long and may weigh over 500 lb! The surface of the shell is fluted and the two valves seal perfectly. Like corals, this gigantic animal, its soft parts vividly coloured, has zooxanthellae living symbiotically in its tissues and actually feeds on them. Colonies of the algae live just under the skin near the margins of the shell, where special light-attracting organs evidently stimulate their growth. The clam is generally found in the larger hollows of coral or on the ocean bed where it remains completely motionless.

The giant clam may represent a serious danger and there have been a number of reports of underwater swimmers who, having accidentally caught an arm or leg between the shell portions of the enormous molluscs, have been trapped and drowned. Although the animal is certainly strong enough to cause accidents, such reports lack confirmation. It is indeed rather unlikely that human deaths have occurred in this way for two reasons – firstly, because the brilliant colours of the flesh are easy to see; and secondly, because the shell portions close very slowly, allowing plenty of time for escape.

Fishes that give and take

The multicoloured fishes of the coral reef find both refuge and protection in their strange, petrified underwater world. Many of them, such as the parrotfishes, feed directly on the coral polyps, nipping them off with their specialised teeth.

But since there are no two species that feed in precisely the same manner or in exactly the same place, there is no competition among these predators. Alongside them live other species which capture the countless invertebrates that make up an important part of this complex marine community.

Among the most specialised fishes of the coral reef are a large number, extremely diverse in shape and colour, which find their food by exploring small cracks in the coral where all manner of tiny invertebrates seek shelter. The characteristic feature of these fishes, no matter which family they represent, is a long beak-like snout, especially prominent in the case of the various butterflyfishes. This dependence on invertebrates is so marked that certain fishes cannot be kept in an aquarium because they refuse to feed on anything else. Konrad Lorenz came up against this problem when he tried to rear the beautiful Moorish idol (*Zanclus cornutus*) in captivity. After several unsuccessful attempts he finally strewed small pieces

Coral fishes form associations of various kinds. Some are mutually advantageous, as in the case of the cleaner wrasse and the grouper (*left*). Similarly, the little cardinal fishes (*right*) find protection among the sharp spines of the sea urchins of the genus *Diadema*, giving a cleaning service to their hosts in return. But the trumpetfish (*centre*) simply makes use of another fish as a kind of decoy to attract the tinier species that constitute its food.

The distinctively marked clownfish lives among the tentacles of dangerous sea anemones, evidently immune to the latter's poisonous stings. Other fishes do not normally venture so close and consequently the clownfish is assured of protection.

The coral reef is an ideal place for the study of animal behaviour. Among many interesting spectacles are the territorial combats of rival butterflyfishes (1) which cross long snouts and push each other with the upper parts of the head, and of clownfishes (2), each trying to strike the adversary's snout with its pectoral fins.

CLOWN OR ANEMONE FISH
(*Amphiprion percula*)

Class: Osteichthyes
Order: Perciformes
Family: Pomacentridae
Total length: up to 6 inches (15 cm)

Easily identified by four broad, transverse, bright orange bands, the first covering the head, the last the tail, separated by three narrower white bands, the largest of which (in the centre) is just behind the pectoral fins. The fins are edged with black with a nearly transparent border.

of winkle flesh over the stones of the aquarium, thus preventing the fish dying of starvation.

The trumpetfish (*Aulostomus maculatus*) has an even stranger habit of lying along the back of another fish while the latter is feeding. Smaller fishes, attracted by the food and unable to see the slender trumpetfish lying in ambush, are swiftly and suddenly attacked and eaten by the predator.

The method of feeding chosen by the cleaner wrasse (*Labroides dimidiatus*) reminds one of the relationship between the African oxpeckers and the various wild and domestic animals on which they perch in order to remove ticks. These fishes rid larger predatory species of parasites and scraps of dead skin. Their coloration is conspicuous—bright blue with a shiny black stripe along the flanks—and they 'introduce' themselves to their hosts by performing a kind of wriggling dance. Aware of their presence, the predators, who will normally attack any species that come close, stay absolutely still and allow the wrasses to roam all over their body, even to the extent of entering the mouth and the gill cavities and mouth. Both parties to this curious association derive some benefit, the wrasse obtaining food, the larger fish relief from organisms that could be damaging to its health.

But that is not the end of the story, for there is a species of blenny (*Aspidontus taeniatus*) or 'false cleaner' fish which deceives the predator into thinking that it is the genuine article. It is able to do this by a trick of imitation, having acquired colours and stripes amazingly like those of the cleaner wrasse and even going through the same introductory 'dance' motions. The predator confidently lets the smaller fish go to work, but instead of carrying out the hygienic ritual the imposter simply tears off a section of gill or fin.

Perhaps the most astonishing association in the realm of the coral reef is that of the clown or anemone fish (*Amphiprion percula*) and certain large species of sea anemone. The latter possess tentacles equipped with stinging-cells, capable of killing any similarly sized animal; but the clownfish swims freely among the flower-like tufts, thereby finding complete protection from predators. Many theories have been advanced as to how it remains unharmed by the venom of the anemone, one of the likelier suggestions being that the fish's skin secretes a substance capable of counteracting or inhibiting the poisons of the stinging-cells.

The tiny cardinal fish (*Siphamia versicolor*) seeks shelter in a similar manner among the long, sharp, dangerous spines of the sea urchins belonging to the genus *Diadema*, possibly cleansing its hosts of parasites.

Dangers of the coral reef

In this underwater kingdom beauty can sometimes be deceptive. Many animals have acquired defensive mechanisms which may spell danger even to man. Thus among the corals themselves those of the genus *Millepora* can deal out painful stings, the consequences of which vary according to the constitution of the victim. The above-mentioned *Diadema* sea urchins often form dense clusters and although their sting is not normally fatal, because non-venomous, they can cause serious injuries, for the rapier-like points penetrate the flesh and then break, so that they are extremely painful and difficult to extract. The bristle worms of the genus *Eurythoe* may be dangerous, but the most unpleasant invertebrates are the cone shells of the genus *Conus*, gastropods with a radula modified to form teeth capable of injecting venom (apparently containing properties similar to curare) into their victims. These molluscs have been responsible for many accidents and have certainly cost the lives of some shell collectors.

Dangerous fishes come in all sizes, ranging from sharks, barracudas, moray eels and electric rays to the much smaller members of the family Scorpaenidae (the colourful lionfishes of the genus *Pterois*, with their long poisonous tentacles, are typical examples) and the stonefish (*Synanceja verrucosa*). The last-named is at all costs to be avoided. Not only does it possess an extremely potent type of venom, capable of killing a man within two hours, but its shape and colour so resemble those of a small rock that it is only too easy for a bather to step on it accidentally.

Among a number of venomous inhabitants of the coral reef are certain brightly coloured molluscs known as cone shells (*above*) and the bizarre-looking lionfish (*below*).

LIONFISH
(*Pterois volitans*)

Class: Osteichthyes
Order: Scorpeniformes
Family: Scorpaenidae

Rounded body with long, tentacle-like fins with separate rays that spread out to make the fish look bigger than it really is. Venom gland at base of each ray of dorsal fin. The strange, beautiful pattern of the fish is made up of a mixture of white and red stripes of variable size.

CHAPTER 7

The frozen wastes of Antarctica

Antarctica is the coldest, remotest and least populated of the world's continents, and the last to be discovered and explored by man. The storm-tossed seas surrounding it and the gigantic walls of ice guarding the interior form a redoubtable double barrier against which many expeditions have foundered but which, for more than a century, have represented a challenge to human enterprise and bravery.

During the 19th century several explorers (recalling the much older tales of a mysterious southern continent) attempted to verify its existence; but it was probably sighted first by an American seal-hunter, Captain Nathaniel Palmer, in 1820. The following year another sealer, John Davis, is thought to have set foot on the mainland. The first explorer to winter on the continent was the Norwegian whaler, Carsten Borchgrevink, in 1900, and a year later three expeditions—British, German and Swedish—were organised. In 1908 the Irish-born Ernest Shackleton set out with a strong team to reach the South Pole but failed to get nearer than 110 miles to his objective. Another member of the party, however, Professor Edgeworth David, climbed Mount Erebus and sledged to the magnetic pole the following summer.

On 20th October 1911 the Norwegian explorer, Roald Amundsen, having received news of Peary's discovery of the North Pole, made a further attempt on the South Pole, reaching it on 14th December of that year. Climatic conditions had been favourable and Amundsen had sensibly laid in ample supplies of food, taking along a team of huskies which, had the need arisen, he and his four companions could have eaten. On the other hand, the British expedition which had set out at around the same time, under the leadership of Captain Robert Scott, ended in tragedy.

Facing page: Probably the best known inhabitants of the Antarctic are the penguins, with their smart black and white plumage.

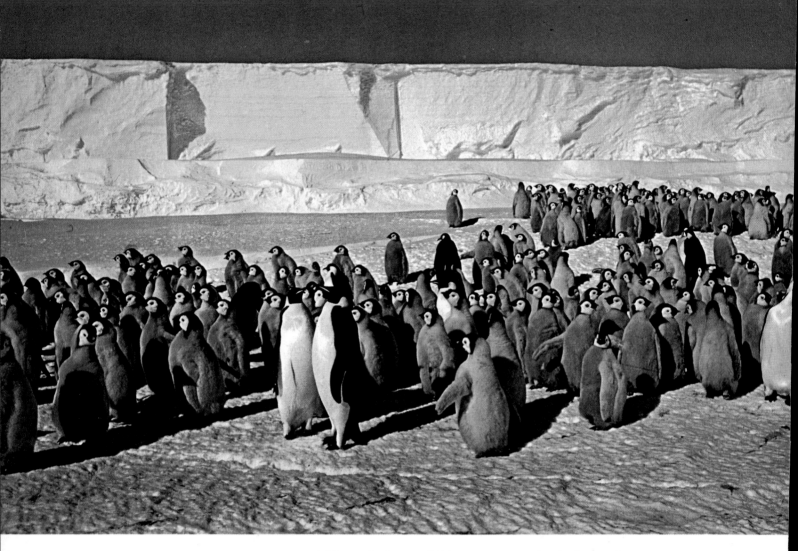

Unlike other Antarctic birds, emperor penguins do not migrate in the winter, for it is during this difficult, dark period that they hatch and rear their young. When they have grown sufficiently, the latter group together under the surveillance of a few adults while the rest of the colony go off in search of food.

Although he and his party reached the Pole on 17th January 1912 the great disappointment of having been forestalled by Amundsen was bitter. Without transportation, assailed by frightful weather and with their food supplies exhausted, Scott and his companions died of hunger and cold only ten miles or so from a food depot, the last date in his diary being 29th March 1912. Eight months later the explorers' bodies were found by a relief expedition and the brave, tragic story was pieced together. Scott's Antarctic bases were retained intact as a monument to the courage and endurance of these and other early heroes of the frozen continent.

Animals of the southern continent

Today a number of countries have a vested interest in the Antarctic region, with permanent bases where scientists can work in tolerable, if not perfect, conditions. Thus man has become the only land vertebrate to inhabit these bleak wastes, but only thanks to the marvels of technology. The birds and seals of Antarctica cannot really be regarded as true land animals because they are entirely dependent for their existence on the ocean and only venture onto the ice to breed.

The comparative poverty of the Antarctic fauna is a consequence of the rarity of plants which constitute the first link in any food chain. In Antarctica there are only three known species of grass and even these are not found in abundance. The other plants consist principally of lichens and mosses, but because their growth is extremely slow they are not sufficient to permit a large animal community to flourish.

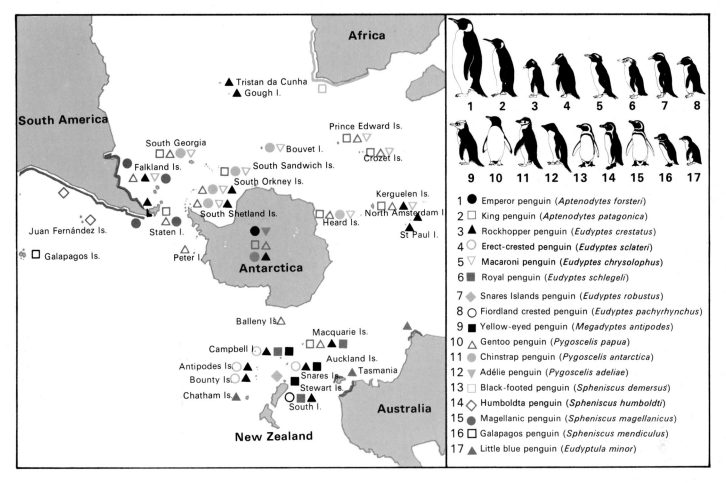

Geographical distribution and breeding areas of the seventeen species of penguins.

The scarcity of plants in Antarctica is the result of the absence of a suitable substratum for rooting, intense cold and extreme dryness. Over a total land area of approximately 5 million square miles, only the mountain summits and certain parts of the coast (probably about 3,000 square miles) are not imprisoned in a thick layer of ice. Certain plants are capable, nevertheless, of withstanding the extreme rigours of the climate, their growth coinciding with that short period when the temperature rises slightly above freezing point. This, however, represents a very small proportion of each year.

Drought is another important factor which restricts plant growth. The part of the continent which receives the highest degree of precipitation—the Antarctic peninsula—does not see more than about 20 inches per year, and much of this falls as snow. In fact plants are able to make use of only a very minute proportion of the water released by snow because this evaporates very quickly under the influence of the violent winds that sweep the continent.

Invertebrates have been rather more successful than vertebrates in establishing themselves in Antarctica, but even in their case species are not numerous. In all there are about fifty known species of insects and mites, the largest of which is a fly (*Belgica antarctica*) which measures barely one-quarter of an inch. Almost half of these species are parasites of birds and mammals which provide them with a favourable microclimate. The rest live principally on lichen and algae which in their turn derive benefit from the enormous quantities of excrement left in the bird colonies.

If one were to judge Antarctica only by the animal communities of its land areas, the overwhelming impression would be of an immense desert, and in fact geographers rank it as such, in the sense of having a low level of precipitation. But the animal communities, by their very simplicity, are of great interest to zoologists. They are comparatively easy to study, especially the ways in which they adapt to cold and drought.

Life in Antarctic seas

In contrast to the relative poverty of animal life on land, Antarctica as a whole, which includes the surrounding oceans, harbours some of the most flourishing marine communities in the world. The basis of this prosperity is the wealth of primary producers, for the currents continuously bring to the surface vast quantities of nutrients from the deeper layers and these are utilised by planktonic algae.

Despite the fact that there are comparatively few species able to adapt to the rigours of the southern oceans, the populations that do exist are very abundant. Among them is one which occupies a key position in all the marine food chains. Science describes this species as *Euphausia superba*, and whalers call it krill. The animal concerned is a small prawn-like crustacean that feeds on microscopic algae and is in turn eaten by fishes, birds and mammals alike. It is undoubtedly because of the preponderance of krill that the largest concentration of whales in the world is to be found in these oceans, including the blue whale, largest mammal that has ever lived. Until recently the blue whale was threatened with extinction. Now officially protected, it is believed to be safe. It is reckoned, nevertheless, that the total whale population of Antarctica has been reduced, as a result of the hitherto uncontrolled activities of whalers, to about one-tenth of its original size.

The enormous cetaceans are not, however, permanent inhabitants of southern seas. In the course of the year they travel both northward and south again. These movements cannot be compared with the mass migrations of birds, in which all members of a species in a given area take their departure simultaneously. In the case of whales it is a gradual process whereby the animals swim south as summer approaches, so that maximum population density is reached in February, and return north at any time between April and July, by which period numbers remaining in Antarctic waters are at their lowest level.

Apart from whales and man, the only group of mammals that has successfully conquered Antarctica is the seals, represented here by four species, each occupying a well defined ecological niche. The Weddell seal is to be found in the most southerly latitudes (regions uninhabited by other animals), not even moving northward in the depths of the Antarctic winter. To withstand the exceptionally bitter cold this seal spends most of its time in water (which is warmer than the air) feeding in the depths on fishes and squids.

The Ross seal is an inhabitant of the ice-floes close to the continental land mass and feeds principally on squids caught under the ice. The crabeater seal (most abundant of these species) lives on krill, its incisors and canines being modified in such a manner that they constitute an efficient filtering

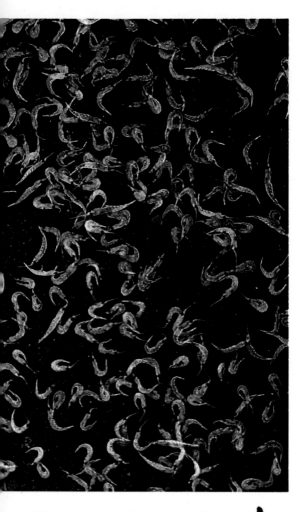

The crustaceans known as krill constitute an essential element in the food chains of the Antarctic seas. Some animals, including certain whales, feed directly on these prawns.

mechanism for capturing the tiny crustaceans that teem in such quantities around these slabs of floating ice.

The leopard seal often visits the fringes of the continent to coincide with the migrations of Adélie penguins. The edges of the ice-pack are the favourite hunting grounds of these aggressive seals whose population is today estimated at about 150,000. Although primarily a fish-eater it occasionally attacks other seals and is also noted for hunting penguins. The latter, as soon as they become aware of the leopard seals' presence, abandon their normally placid activities and make off at top speed, leaping up to 12 feet in the air to escape their predators. They seem to be diving the wrong way up as they explosively break surface to pass from one medium to another.

Antarctic sea birds

The most familiar inhabitants of Antarctica are undoubtedly the birds. Among them are two species which, unusually, obtain all or a large part of their food out of water. The first of these, and the only one not possessing webbed feet, is the sheathbill, which generally roams around the nests of other sea birds, snatching up any remains of food. The other exception to the rule is the great skua, equally at home on land, with a broadly based diet which consists mainly of fish with a large proportion of carrion as well as eggs and penguin chicks.

The best known birds, however, are the various species of albatrosses, petrels, terns, cormorants and penguins. These can roughly be divided into two groups, those that breed on islands and those that breed on the mainland proper. Only about 30 or so species belong to the latter group. To be quite accurate, not even the popular penguins (so frequently photographed against an icy background) are exclusively polar birds. They are, in fact, birds of the southern hemisphere which happen to have colonised regions where no predatory land mammals roam. Although, by and large, they all have a southerly distribution, the Galapagos penguin is to be found as far north as the equator. Among other species, only the emperor and Adélie penguins are exclusive to the continent of Antarctica.

All these birds find plenty of food in the sea and for that reason their populations are extremely large. But there are not all that many suitable nesting sites and on those rocky islets where conditions are right, colonies of different species are often seen mingling with one another.

The world of penguins

The oldest eye-witness description of penguins is probably that of the French explorer Beaulieu. In 1620 he came across them off the southern tip of Africa and spoke of them, not very perceptively, as 'feathered fishes'. It is likely that he would have identified them more accurately as birds had he seen them on land, but perhaps he may be forgiven his 'howler' for there are few birds anywhere in the world so magnificently adapted to their marine surroundings.

The babies of most penguin species have little protection from the elements because the adults build their rudimentary nest on the ground, at best in the shelter of a rock. But the single chick of the emperor penguin (*top*) is kept warm, covered by a flap of abdominal skin, between the feet of either parent.

In spring the female Adélie penguin lays her eggs in a nest composed of pebbles, then wanders away to the distant sea to feed, leaving her partner to incubate them.

ADÉLIE PENGUIN
(*Pygoscelis adeliae*)

Class: Aves
Order: Spheresciformes
Family: Spheniscidae
Height: 28–30 inches (70–75 cm)
Wing-length: 7½ inches (18 cm)
Weight: 6½–13 lb (3–6 kg)
Diet: basically krill, some fishes
Number of eggs: 2
Incubation: about 35 days

Species forms huge colonies on coasts of Antarctica and rocky islands. In adults, head, back and upper sides of wings are black, neck, chest and belly white. Newborn chick covered with grey down.

People who have seen penguins only in zoos—a completely artificial environment—tend to regard them with rather condescending amusement, but can have no idea of their vastly different behaviour in the wild. Concentrated, as the majority of these birds are, well away from temperate latitudes, it is only thanks to film and television cameras that those of us who are not scientists can begin to appreciate their astonishing adaptability. What perhaps cannot possibly be realised is how abundant they are (except for a few species) and what enormous colonies they form. On one of the South Sandwich Islands, for example, there are some ten million chinstrap penguins (*Pygoscelis antarctica*). About half a million royal penguins (*Eudyptes schlegeli*) live in the Macquarie Islands; and the odour of the half million rockhopper penguins (*Eudyptes crestatus*) that nest on Nightingale Island is something that no visitor can quickly forget. The largest and undoubtedly the most beautiful species is the emperor penguin (*Aptenodytes forsteri*), of which 100,000 individuals are massed on Coulman Island.

Superficially, most penguins (apart from the large emperor and king penguins) look much alike, with their characteristic black and white livery, upright stance and comic waddling gait. But, apart from dimensions, each species has a distinct facial pattern which makes for easy identification.

Swimmers extraordinary

The unmistakable features of the penguin—erect posture, swaying gait, thick plumage, heavy body and short wings—are all adaptations to aquatic life. Because of this high degree of specialisation the anatomical and physiological modifications exhibited by the different species are more pronounced than in any other groups of water-dependent birds, including ducks and cormorants.

It would be interesting to know much more than we do about the evolutionary history of penguins and to have indisputable evidence of the various stages of their development from primitive times to the present day. What fossils have been found are of fairly recent geological origin. The oldest, discovered in New Zealand, do not date back more than some 40 or 50 million years. They belonged to individuals whose feet and wings were already much like those of modern penguins, so that they cast little fresh light on the subject. Ornithologists are convinced that penguins are descended from sea birds and that at some stage of their history they went through a phase similar to that of the auks, gradually losing their flying powers as their body weight increased. This is of course what happened to the great auk of the North Atlantic which, as a result of this handicap, was exterminated.

Even at the time when penguins must have been capable of flying they would have had to contend with the competition of other sea birds, and under such pressures they began to specialise in diving and swimming far under water for progressively longer periods. The positioning of their feet well to the rear of the body is an obvious advantage for swimming; and the thick

layer of fat enveloping the streamlined body has important insulating properties, helping to keep their temperature stable and to avert excessive energy loss. Furthermore, long wing feathers, vital to all birds whose element is the air, are obstacles to movement through the water. Most diving birds keep the wings folded when submerged and propel themselves by means of their webbed feet, which act as paddles. But penguins move their wings when swimming, as if they were flying in the water. The mechanics of this form of locomotion have necessitated a transformation of the forelimbs into flippers. Some such modification is in evidence among the Alcidae (auks) as well, but only in the case of the Spheniscidae (penguins) has it reached such a pitch of perfection.

Paradoxical as it may seem, the fatty layer of tissue is such an effective insulating medium that the main problem for penguins, especially those living in rather warmer regions, and particularly when out of water, is to keep the body cool. Unlike most other birds (including tropical species) penguins have feathers growing uniformly all over the body, lacking those naked patches which help to release heat accumulated as a result of ordinary muscular effort.

Dr Bernard Stonehouse, formerly of the University of Canterbury, New Zealand, who has studied the penguins and other animals of Antarctica, has compared this covering of feathers and adipose tissue to a heavy, waterproof overcoat. It enables a penguin to spend days, even weeks, in the sea; yet it does not handicap the bird unduly as it goes about its vital activities, such as swimming rapidly in pursuit of prey, running about on dry land, engaging in ritual combat, and reproducing. One way in which the bird can lose heat is by gently ruffling its

To satisfy the enormous appetites of their young, Adélie penguins have to spend much of their time fishing. Each pair only take responsibility for their own offspring and should anything happen to them the orphans will probably die of hunger.

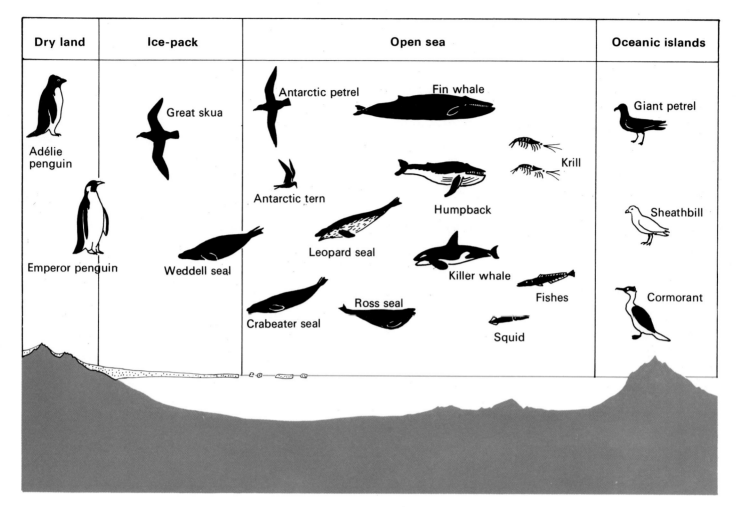

Among the birds of Antarctica some, such as the Adélie and emperor penguins, are perfectly adapted to withstand the harsh climate. Others, including skuas and terns, visit the continent only in summer and fly northward in the winter, spending the season on the open sea. These islands, with their more moderate climate, are convenient breeding sites for a number of species, such as cormorants, which are never seen around the coasts of Antarctica. As for mammals, it is the seals—especially the Weddell seal—which are most attracted to the frozen wastes. The whales, although they move north or south according to the season, are only to be found in the open sea.

feathers; but it has other ways of keeping cool, for the inner edges of the wings and the upper sides of the webbed feet (not enveloped in fat) also help to let out heat.

Out of their element

Penguins often leave the water to rest on the ice or rocky shore; but apart from these brief excursions outside what must be regarded as their natural element, there are two periods in their annual life cycle when they are involuntarily confined to dry land and thus unable to pursue their fishing activities. One of these is during the season devoted to mating, egg-laying and incubation, the other when they shed their feathers.

Around October, when strong winds are still battering the frozen surfaces of the continent, colonies of Adélie penguins abandon the sea and set out on long journeys across the ice to their breeding sites. They may travel as much as 60 miles and when they arrive at their destinations they are likely to find the nesting places still ice-bound. The courtship displays and pairings-off take two or three weeks. As soon as each female lays her two eggs she returns to the sea to feed, leaving her partner to incubate. She may be away for two or three weeks, during which time the male will have lost weight at the rate of one pound per week. Should the weather still be cold the journey back to open water will be long and arduous; and when she makes her way back again (her third journey) she may find the nest empty and abandoned, the eggs having been stolen by skuas. But if the temperature has risen, allowing the sea ice around the colonies to melt, the periods of absence and fasting will be reduced, improving the prospects of hatching.

Whatever the climatic conditions, both adult penguins are compelled to spend a fairly long time out of water so that, after mating, they will live exclusively on accumulated reserves of body fat.

When the chicks are hatched food is once more abundant and both parents fish indefatigably for the family. It is essential that they themselves put on weight before they are again immobilised on land during the moulting period.

These harsh necessities sometimes force the parents to abandon their offspring. Inexperienced individuals which are late in pairing off and mating may find themselves in mid-February with chicks that have not yet acquired feathers and are therefore incapable of becoming independent. Were the parents to continue feeding them they would expend so much effort that when the time came to moult they would still be so thin and weak that they would in all probability die. Faced with such a dilemma, they cease to rear their chicks, leaving them to perish and fishing exclusively for themselves. Thus although they will have lost their brood they are strong enough to survive the critical moulting season.

The stately emperor penguin

When spring, with its life-renewing forces, reaches the northern hemisphere, summer is giving way to autumn in southern latitudes. The days become shorter, the temperature drops, the trees lose their foliage and nature sinks gradually into sleep. In the extreme south, in Antarctica, snowstorms rage with greater force and frequency and the oceans freeze. All the birds which have spent the summer in these parts, feeding in the sea, once more begin to wing their way northward. There is only one exception to this general rule, one species which, following what would appear to be the wrong itinerary, travels in precisely the opposite direction towards the frozen regions that other birds have by now abandoned. This is the emperor penguin, identifiable by its size, its yellowish-white front and its orange-gold 'ear' patches—largest and most handsome representative of the whole penguin tribe.

The southward journey of the communities of emperor penguins is, of course, no error of navigation but quite deliberate, for in contrast to other species these birds actually mate and incubate their eggs in the winter instead of the spring. As they get nearer to their breeding sites individual small groups merge into immense colonies, making an unhurried trek over the ice and reaching their destinations around the end of March or the beginning of April.

Zoologists are uncertain whether emperor penguins make contact with their partners of previous years, as is the case with a high proportion of Adélie penguins, but certainly in the hubbub of noise which accompanies the first phase of the courtship ritual a bird will often assume a watchful, expectant air, as if trying to isolate and identify a familiar voice in a chorus of many thousands.

During May or June each female lays a single egg which her

EMPEROR PENGUIN
(*Aptenodytes forsteri*)

Class: Aves
Order: Sphenisciformes
Family: Spheniscidae
Height: 45–46 inches (114–117 cm)
Wing-length: 13½ inches (34 cm)
Weight: 57–64 lb (26–29 kg)
Diet: fishes, squids, crustaceans
Number of eggs: one
Incubation: 62–64 days

Largest of all penguins. Head, back and upper sides of wings black; belly and lower sides of wings white. Orange mark in ear region. Long, slender bill with lilac mark along edges. Breeding takes place in winter after penguins assemble in large colonies. Newborn chick covered with grey down. Head black, but two white marks around eyes. Adult plumage acquired after two years.

■ *Chionis alba* (breeding zone)
■ *Chionis minor* (breeding zone)
■ *Chionis alba* (wintering zone)

Geographical distribution of the wattled sheathbill (*Chionis alba*) and the lesser sheathbill (*Chionis minor*).

LESSER SHEATHBILL
(*Chionis minor*)

Class: Aves
Order: Charadriiformes
Family: Chionididae
Wing-length: 8¾ inches (22 cm)
Weight: about 1 lb (400–500 g)
Diet: eggs and chicks of penguins and other birds; carrion, molluscs
Number of eggs: 2–4
Incubation: 29 days

Plumage completely white, black bill. Young similar to adults, but newborn chicks have brownish-yellow or grey back, light chestnut forehead, white throat and greyish abdomen. The wattled sheathbill (*Chionis alba*) is distinguished by its yellow beak.

partner immediately balances on his feet and covers with a fold of abdominal skin. While this manoeuvre takes place both birds let out piercing cries and display to each other by pointing the beak to the region of the incubating pouch as if surprised by the event or afraid of losing the egg. But the female soon loses interest in the proceedings, wandering away from her mate to join other females. Soon afterwards they all leave the colony and make their way back to the sea, a journey which is far longer than the previous one because autumn is now over and the ice-sheet has moved many miles farther north. Nevertheless the females set out, quite undeterred, on their travels, their characteristic waddling movements punctuated from time to time with belly slides. Meanwhile the males watch unconcernedly as the outlines of their consorts vanish over the horizon. The heavy responsibility of incubation now rests with them alone.

This incubation period lasts approximately two months, during which time the male eats nothing and has to endure temperatures down to $-40°C$, with violent winds of up to 90 miles per hour. Logic would dictate that under such miserable conditions so many adults would die of exposure that a high proportion of eggs would be lost. Yet this does not happen, for emperor penguins are perfectly adapted to these rigours.

On days when the storms subside and weather conditions improve slightly, the males warily take a few steps forward, warming and preening themselves, carefully balancing the precious egg on each foot so that it does not tumble onto the ice. But when the weather deteriorates they cluster close to one another, not moving for days on end while the tempest continues to rage all around them. This is not for comfort or reassurance but in order to avoid excessive heat losses and to save energy. Bernard Stonehouse, who spent a winter in Antarctica studying a breeding colony of emperor penguins, noted that those individuals which he separated from their companions got thinner much more rapidly than those allowed to remain in groups. It appears that during the incubation period the body temperature of the penguins decreases to some extent, which likewise slows up heat loss. Furthermore, Stonehouse noted that each bird took turns in assuming a position on the outside of the group and that the whole colony would slowly veer round, like a huge fan, according to the direction of the wind, so that each bird kept its back to the storm.

While the males incubate, the females fish in the sea on the fringes of the ice-pack which, as winter goes on, gradually recedes. Approximately two months will have elapsed since egg-laying when they all return to the breeding colony, calling for their partners. When each pair is reunited the male hands over the egg for the final stage of incubation to the female, then waddles off in turn to the sea to find food.

The females do not, however, invariably return on time. They may be caught and killed by a leopard seal, suffer some kind of accident or be delayed by a storm, so that the chick may well hatch while they are absent. In such a case the male is forced to feed the baby without leaving its side, otherwise it would surely die of cold on the ice. In such situations the male shows

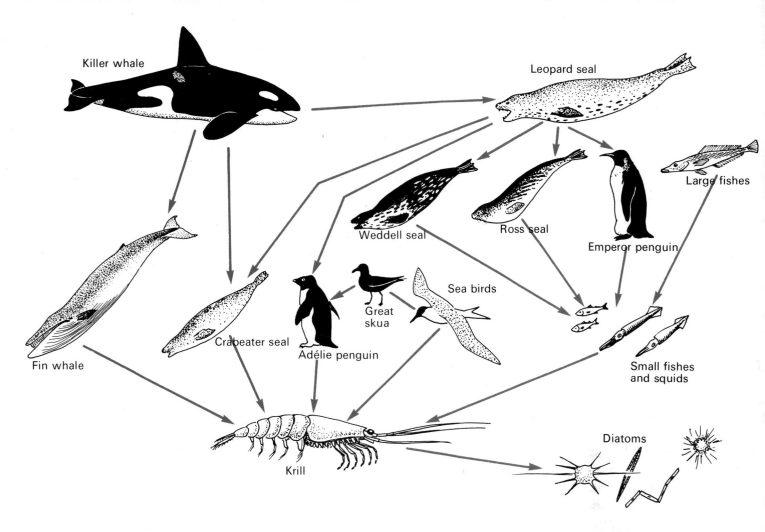

astonishing powers of endurance. Even after fasting for three months he is still capable of regurgitating a little food, so that the baby can survive a few days until its mother arrives. In the majority of cases the female gets back in time to complete incubation and regurgitate food for the chick.

For the first two months of its life the young emperor penguin remains comfortably installed between the feet of one or other parent who take turns to provide it with shelter and to go fishing in nearby waters, now that the winter ice has almost disappeared.

Between July and September (the exact time varies according to colonies) the baby leaves its warm refuge and wanders off to join companions of its own age. At this stage it looks like a grey ball of fluffy down. Large groups of youngsters form, and from time to time each bird strays off to launch an appeal to its parents who soon come to the rescue with food.

By the end of November (the Antarctic spring) the young emperor penguins begin to lose their initial covering of down and within the next couple of months acquire their adult plumage. Having done so they are ready to enter the sea for the first time, where they learn how to fish and lead an independent life. Meantime the adults are building up their reserves of fat prior to moulting, after which they return to the ocean, now teeming with fish.

The opportunistic sheathbills

Survival in Antarctica depends on being able to secure an adequate supply of food from the sea. For vertebrates incapable of fishing, prospects of living permanently in these southern

The importance of krill in the ecosystem of the Antarctic Ocean is shown by this diagram of a variety of food chains. Some are more complex than others but all include krill as a basic element.

Following pages: During the period when they moult their feathers penguins are unable to swim and collect in large groups on the desolate beaches of Antarctica. Since they cannot feed normally they survive on previously accumulated reserves of fat.

climes would be dismal, for there are simply not enough plants and invertebrates on land to sustain them. A higher vertebrate normally has no alternative, therefore, but to be in some way dependent upon the sea. Nevertheless there are a few notable exceptions—birds that have adapted to a style of life that does not require them to procure food directly from the water. The two species of sheathbills are medium-sized birds with white plumage. The wattled sheathbill (*Chionis alba*) has a yellow beak, the lesser sheathbill (*Chionis minor*) a black one. In certain features of skeletal structure they resemble oyster-catchers, in other ways they are similar to gulls. Ornithologists believe that they may be the modern descendants of the common ancestors of all sea birds.

Sheathbills usually settle near colonies of penguins and cormorants for it is upon these birds that they rely for food in spring and summer. They mate in October or November and around the end of the year each female lays two or three (sometimes four) eggs, in a nest made of scraps of plants, feathers and other materials, placed in the shelter of a rock. The chicks hatch after a month, begin to fly seven or eight weeks later and soon afterwards become completely independent.

From the moment the courtship ritual commences until the time when the young leave the nest for good, the sheathbills find their food among the penguin colonies, roving to and fro, ready to snap up any remains. Whenever an adult penguin opens its beak to regurgitate food for its chick, a sheathbill may be hovering nearby in the hope that a morsel will be dropped. It may show even more aggressiveness, harassing the penguin so as to disturb the feeding procedure or forcing it to release its hold on a fish that it may have in its beak. If the opportunity arises, a sheathbill will steal an egg, snatch up a chick or even kill an injured or handicapped adult.

When the young penguins quit the breeding colonies, sheathbills modify their feeding habits, congregating around the coasts and pecking through the seaweed for dead fishes and sea invertebrates. When winter comes they migrate northward as far as Tierra del Fuego, Argentina and the Falkland Islands. Here they continue to feed along the shores while the young make for inhabited areas, in the expectation of picking up various forms of waste matter.

The marauding skuas

From October onwards flocks of dark brown, hook-billed birds, rather smaller than herring gulls, converge on Antarctic waters, having spent the past six months scattered across the seas of the southern hemisphere. These are great skuas (*Catharacta skua*) returning to breeding sites on the coasts of Antarctica, the surrounding oceanic islands and the southernmost tip of South America. The species is also found in the northern hemisphere, in Iceland, Scotland and neighbouring islands, having presumably originated in the south.

Once arrived in Antarctica, the skuas settle near colonies of penguins, petrels and other birds on which they will feed during

---

GREAT SKUA
(*Catharacta skua*)

Class: Aves
Order: Charadriiformes
Family: Stercorariidae
Length: 22–26 inches (55–65 cm)
Wing-length: 15½–16 inches (39–40 cm)
Weight: 2¾–3½ lb (1·3–1·6 kg)
Diet: fish, eggs, chicks and adult birds, carrion, marine invertebrates, refuse
Number of eggs: 2
Incubation: 28–32 days

Looks like a large dark gull. White mark at base of primary remiges. Bill and feet black. Chick covered with grey down. Plumage of young slightly darker than that of adult.

the breeding season. In November or December each female lays one or two eggs in a pebble- or moss-lined hollow, incubating them for about a month. The chicks stay put for six or seven weeks, then leave the area to wing their way far out to sea, returning some three or four years later as fully fledged adults.

While on land skuas are fiercely territorial, patrolling their hunting grounds and diving on any intruders. When at sea they will feed on marine animals but when incubating their eggs and raising their young they are true marauders, the terror of all other birds. They make audacious attacks on penguin colonies, stealing eggs and devouring chicks, and will in fact feed on the body of any dead animal. So ferocious are they that even a member of the same species, should it suffer injury, will flee a healthy individual for fear of attack. Like sheathbills, skuas are accustomed to living near human settlements, making the most of any available waste matter.

The great skua plays the role of super-predator in the ecological pattern of Antarctica where other carnivores are few and far between; but in winter they abandon these shores and lead a pelagic life similar to that of other sea birds.

During the summer sheathbills (*left*) and the great skua (*right*) settle along the coasts of Antarctica, raiding the colonies of penguins and other sea birds. They are both predators and scavengers.

ORDER: Sphenisciformes

The order Sphenisciformes includes particular sea birds that are incapable of flying. Exclusive to the southern hemisphere, they are known collectively as penguins.

Perfectly adapted to an aquatic existence, penguins are considered by ornithologists to be most closely related to the Procellariiformes (albatrosses, fulmars, petrels, shearwaters and diving petrels).

The short feet are placed well to the rear of the body and possess four toes. Some penguins waddle rather clumsily over solid ground, others hop; sometimes they slide on their belly over the ice, propelling themselves with wings and feet.

The forelimbs have been converted into stiff flippers which move freely from the shoulder and which are used for locomotion both in the water and on the ice. The short tail consists of rigid feathers. Contrary to most birds, the body feathers grow uniformly all over the body, without intervening naked patches, so that little heat is lost.

Penguins are highly gregarious birds, especially during the breeding season when they form enormous colonies. Apart from those of the genus *Aptenodytes* (which balance their single egg on their feet and cover it with a fold of skin in order to incubate) the various species usually make a rudimentary nest in a hollow on the ground, and line it with stones or grass. The females lay two, sometimes three, eggs.

In all species the front part of the body is white or yellowish and the back black, without any differences in coloration between male and female. Nevertheless, each species is distinguished by the markings on or around face and neck, or even by the presence of a crest. There is considerable variation in size as well, the smallest species standing about 16 inches tall and the largest around 4 feet.

There is only one family (Spheniscidae) which contains six genera and seventeen species.

The genus *Aptenodytes* consists of two species, the emperor penguin (*Aptenodytes forsteri*) and the king penguin (*Aptenodytes patagonica*). These are the largest of living penguins, the former measuring up to 48 inches, the latter about 38 inches. Both are distinguished too by yellow or orange patches in the ear region.

The three representatives of the genus *Pygoscelis* are the Adélie penguin (*Pygoscelis adeliae*), the gentoo penguin (*Pygoscelis papua*) and the chinstrap penguin (*Pygoscelis antarctica*). These are the next largest, about 30 inches tall. Like the emperor and king penguins, these birds are typical of the Antarctic mainland and surrounding islands.

There are six species belonging to the genus *Eudyptes*, the so-called crested penguins. They are the fiordland crested penguin (*Eudyptes pachyrhynchus*), the erect-crested penguin (*Eudyptes sclateri*), the rockhopper penguin (*Eudyptes crestatus*), the macaroni penguin (*Eudyptes chrysolophus*), the royal penguin (*Eudyptes schlegeli*) and the Snares Islands penguin (*Eudyptes robustus*). All are inhabitants of New Zealand and surrounding islands, and this is also the habitat of the only species of the genus *Megadyptes*, the yellow-eyed penguin (*Megadyptes antipodes*).

The genus *Eudyptula* is represented by the little or little blue penguin (*Eudyptula minor*), smallest of all penguins, which breeds in Australia and New Zealand. Finally there are the four members of the genus *Spheniscus*—the black-footed or jackass penguin (*Spheniscus demersus*), which breeds in South Africa, the Peruvian or Humboldt penguin (*Spheniscus humboldti*) and the Magellanic penguin (*Spheniscus magellanicus*) which are found off the tip of South America, and the Galapagos penguin (*Spheniscus mendiculus*) which, as its name indicates, is an inhabitant of the Galapagos Islands.

Facing page: Few animals have been able to adapt to the bitter climatic conditions of the Antarctic, but among them are certain species of penguins. Those that have established themselves are to be found in enormous colonies, as is shown in this photograph of king penguins.

CHAPTER 8

Life on the Continental Shelf

Thanks to recent developments in underwater photography, many marvels of the world beneath the sea have now been revealed to readers of natural history books, cinema audiences and television viewers in their millions. The cameras have opened our eyes to the secrets of the deep—the strange and colourful forms of the coral reef, fields and forests of marine plants, immense caves and canyons, the ghostly hulks of great ships wrecked centuries ago, and a multitude of unbelievably beautiful fishes.

What perhaps we do not immediately realise, however, is that the only reason that we are privileged to glimpse some of the wonders of this magical world of the sea is that it is bathed in sunlight. In these relatively shallow waters, where the sun's rays penetrate rock cavities, illuminate floating strands of weed and flash like shafts of gold off beds of soft sand, the sea gives the impression of being warm, welcoming and somehow familiar. But where the gentle slope of the ocean bed drops abruptly and steeply downward, the light gradually fades into the cold, terrifying darkness of the abyss; and from this immense, gloomy realm, with its crushing water pressure, man (except when aided by technology) is barred.

The illuminated zone of the ocean, accessible to those equipped with aqualungs, masks, flippers and the like, coincides more or less with the area covered by the continental shelf, which is really the underwater extension of a land mass. In primitive times this terrain, now totally submerged, was caressed by breezes and warmed by the sun; the grooves and furrows left by streams and rivers or etched by eroding winds, waves and storms can clearly be seen. Here on the continental shelf a wealth of plant and animal life is concentrated, making up a

Facing page: Most of the animals living on the continental shelf find refuge in rock cavities. A typical example is one of the larger crustaceans, the crawfish or spiny lobster.

complex ecological scheme which is of absorbing interest both to the adventurous amateur and the professional naturalist.

This easily explored area of ocean, which extends from the tidal zones close to shore down to a depth of about 600 feet where the continental slope begins, has the initial advantage, therefore, of being well lighted. But there are other beneficial factors to be considered. Life is easier here than on the coast, for the rhythm of the tides is hardly perceptible. Thus there is no risk of desiccation and fluctuations in salinity are due only to freshwater streams and rivers flowing into the sea, or to submarine currents. Furthermore, because of its proximity to the coast, the water is particularly rich in nutrients, either in the form of mineral salts or broken down organic substances which are indispensable to the development of marine plants. As a result of plant activity they are transformed into living matter to be used as food by the various animals of the ocean. But whereas on the seacoast the different animal communities are distinct and often separate by reason of zonal subdivisions, the inhabitants of the continental shelf, where conditions are more stable and predictable, cannot be so conveniently classified. It is obvious, however, that the number and variety of species and individuals are directly proportional to the amount of sunlight, the depth of water and the quantity of floating plant and animal material available as food. The farther down one explores, the fewer are the signs of life.

Plants under the sea

The vegetational cover of the continental shelf is in some respects similar to that found on dry land. Nor is the parallel confined to mere appearance for the same biological laws operate both in and out of the water.

The microscopic plants that float near the surface–phytoplankton–are the most important producers of organic matter in the sea. Marine algae provide food and refuge for a wide and varied animal community, in the same manner as land plants.

In the upper part of the continental shelf, an area which is in close contact with the shore, the characteristic vegetation is determined by the nature of the substratum in which it is rooted. In rocky zones near the coast there are vast stretches of strap- or ribbon-like algae known as laminaria, which in some places attain considerable dimensions. Familiar species include tangle (*Laminaria digitata*), cuvie (*Laminaria hyperborea*), and sea belt (*Laminaria saccharina*). They form broad barriers offshore, sheltering many animal communities in the vicinity from the fury of the waves. In fact the influence of laminaria on the effects of tides and the movement of water is so pronounced that they have been introduced in some coastal regions both to protect exposed beaches and to facilitate the reproduction of certain marine animals. In areas where the composition of the sea bed is unsuitable for natural growth, experiments have been made with plastic imitations of the seaweed.

These submarine prairies are not composed entirely of laminaria. Other typical forms of algae around the coastal fringe include those of the genera *Nereocystis*, *Macrocystis*, *Porphyra* and *Caulerpa*, which are nowadays commercially exploited by the food and chemical industries.

Many zoological groups find in these fields of seaweed both shelter and protection. Some, particularly the invertebrates, use the algae as a support for attachment, others use the root-like filaments for camouflage. There are fishes whose colours and markings match this background to perfection; and bryozoans such as the sea mosses cover the algae with a delicate, net-like tracery. Apart from these there are animals which feed exclusively on algae, including sea hares of the genus *Aplysia* and sea urchins of the genera *Echinus*, *Paracentrotus*, etc. All

The parts of the continental shelf nearest to shore are characterised by immense fields of algae, especially those known collectively as laminaria, with ribbon-like laminae or fronds. This species is commonly called tangle.

Many plant-eating animals live on the ocean bed in the photic zone where sunlight freely penetrates. The picture on the left shows a sea urchin of the species *Echinus melo*. On the right are sea fans (*Eunicella*), sea squirts (*Halocynthia*) and a sea urchin of the species *Paracentrotus lividus*.

these species (which can be regarded as true marine phytophages) have the same type of impact on sea vegetation as herbivorous mammals have on the plants of woodland and savannah.

At deeper levels where the substratum is soft there are phanerogams (seed plants) which form fields of vegetation far more extensive than those of algae and which are amazingly adapted to saline conditions. They belong to the families Hydrocharidaceae and Potamogetonaceae. Among the most important plants of the former family are those of the genus *Thalassia*, found in vast quantities on muddy sea beds or in smaller patches where the substratum is firmer and more compact; and, representing the latter family, the genera *Posidonia* and *Cymodocea* (almost exclusively Mediterranean) and *Zostera*, characteristic of the North Atlantic.

These immense underwater plantations likewise provide places of refuge for innumerable animals and plants so that in their entirety they can be regarded as genuine ecosystems. It is not uncommon for such communities to include algae (*Ulva*, *Fossiella*, etc), crust-forming animals similar to those found along the coasts (polyps, gastropods, bryozoans and sea stars), as well as species that are dependent on this type of substratum for food and domicile, such as sea cucumbers, certain bivalve molluscs and tube worms.

The continental shelf is the realm of burrowing animals. Their bodies are especially adapted for this purpose and there are few substances through which one or the other cannot bore. Some of them are barely out of the larval stage before digging a hole in the sea bed and this refuge gradually increases in size as they grow. The burrow's only link with the exterior is a small opening through which food can be extracted, so that when the animal is fully adult it will be effectively protected both from the waves and from potential enemies.

Some of these animals, secure in their burrows, create a great deal of damage. Thus certain sponges (*Cliona*), acorn shells (*Lithotryta*) and necklace shells (*Natica*) attack beds of oysters and other molluscs. When numerous they are capable of destroying an entire breeding area.

Even more destructive are the shipworms (*Teredo*). These worm-like bivalve molluscs have a small shell situated on the extreme front part of the body with which they rasp away wood. A striking example of their capacity for causing wide-scale damage is recorded from Holland where, in 1730, dikes and sluices built to protect the coast of the Netherlands were so undermined by the activities of shipworms that their woodwork literally crumbled under the impact of the North Sea waves,

Perhaps the most specialised predators of hydroid polyps are the molluscs known as sea slugs. The polyps serve as a means of defence as well as food, for those stinging cells which are not digested accumulate in the sea slug's dorsal papillae.

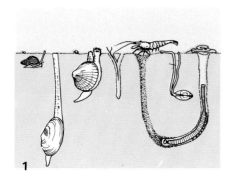
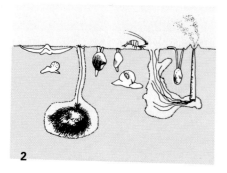

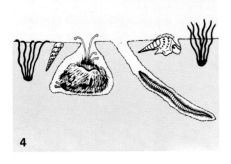

Regardless of geographical location, similar groupings of marine animals are to be found on the floor of the continental shelf at given depths and on the same types of substratum. These drawings show some of the principal species in communities of *Macoma* (1), *Venus* (2), *Syndosimya* (3) and *Amphiura* (4).

letting the water through in many places and flooding large areas of land. Some scientists hold shipworms partially responsible for the disaster that overtook Philip II's Invincible Armada in 1588. They point to the fact that the Spanish ships remained at anchor in Lisbon, prior to the attempted invasion of England, long enough to give the molluscs ample time to gnaw away at their hulls. When the galleons were caught in storms around the British coasts, they foundered.

The bed or benthic zone of the continental shelf is very varied. It is noticeable, however, that at a certain depth and given the same type of substratum, the animal inhabitants of this zone show an extraordinary resemblance to one another, no matter what the geographical location. According to Gunnar Thorson, it is possible to distinguish clearly defined groups of marine animals, the constitution of such communities being mainly determined by the depth of the sea and nature of the substratum.

Fairly shallow waters (30-200 feet) close to river estuaries are populated by *Macoma* communities, composed principally of bivalve molluscs such as the Baltic tellins (*Macoma*), softgapers (*Mya*) and cockles (*Cardium*), as well as lugworms (*Arenicola*). At lesser depths, close to sandy beaches, the most typical forms of marine life are tellins (*Tellina*), wedge shells (*Donax*), artemis (*Dosinia*) and sea stars (*Astropecten*).

The so-called *Venus* communities are found in the open sea and on sandy bottoms ranging in depth from 25 to 140 feet. These comprise venus shells (*Venus*), small clams (*Mactra*), tellins (*Tellina*) and lantern shells (*Thracia*), as well as heart urchins (*Echinocardium* and *Spatangus*), bristle worms (*Ophelia*) and a number of sea stars.

The *Amphiura* community, which flourishes in muddy sea beds up to a depth of about 325 feet, includes tusk shells (*Dentalium*), sea pens (*Pennatula* and *Virgularia*), bristle worms (*Nepthys* and *Terebella*), tower shells (*Turritella*) and burrowing brittle stars (*Amphiura*). Also found on muddy bottoms but in shallower water rich in organic substances are communities of *Syndosimya*–lamellibranch bivalves–together with the related basket-shells (*Corbula*) and nut shells (*Nucula*), bristle worms (*Nephthys* and *Pectinaria*) and sea urchins (*Echinocardium*).

A number of crustaceans mingle with these marine communities, among them spiny lobsters (*Palinurus*), lobsters (*Homarus*), squat lobsters (*Scyllarus*) and others belonging to the Reptantia group. These all possess a fairly long tail and although their customary way of getting about is by walking along the bottom, they do make brief swimming excursions. Other crustaceans, however, with short tails that are of little or no propulsive value, are more sedentary by habit. These include spider crabs (*Maia* and *Lissa*) and edible crabs (*Cancer*).

Other crustaceans which spend a part of their life on the sea bed (although, by reason of their swimming abilities, they also mingle with the animal communities of open water zones) include the shrimps of the genera *Crangon* and *Leander*, as well as various free-swimming crustaceans such as the shrimps of the family Penaeidae.

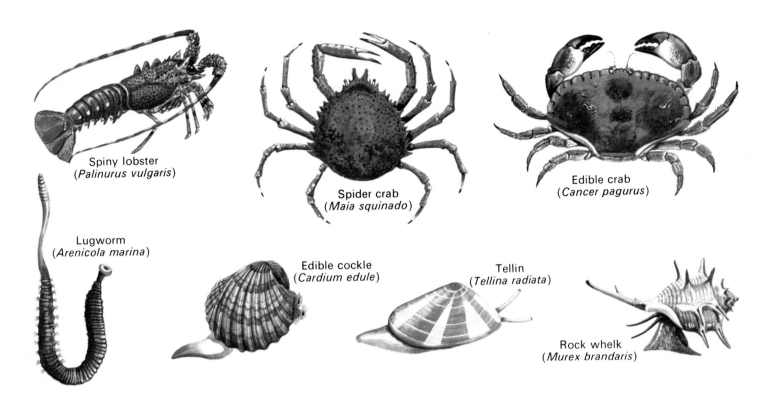

A selection of invertebrates typical of the continental shelf.

Life and death on the continental shelf

The continental shelf corresponds more or less to the so-called photic zone, that area of ocean which receives sufficient sunlight for plant growth. The wealth of life in this region is due principally to that solar irradiation and to the abundant presence in these waters of mineral salts.

As previously mentioned, inorganic matter in the sea is converted into organic material by plants as an adjuct to photosynthesis, for which sunlight is essential. The primary producers responsible for photosynthesis—phytoplankton and benthic plants—are consumed by the plant-eating crustaceans known as zooplankton (copepods and euphausiaceans. It is worth stressing that at first glance there might appear to be a disproportionate balance of phytoplankton and zooplankton, since investigations have shown that the mass of zooplankton is between one and a half and two times as great as that of its plant food. The reason for this is that the rates of reproduction of marine plants and animals are very different. Although phytoplankton is consumed almost as soon as it appears, it is renewed by virtue of a very rapid life cycle (one or two weeks for each generation), whereas the process is much slower for the zooplankton.

Zooplankton, in its turn, constitutes the basic food of a large number of pelagic species, ranging from small crustaceans and fishes to large whales. But planktonic animals also have other less well-known predators, some of which are particularly voracious, notably the arrow-worms of the genus *Sagitta*, which derive their name from their long, arrow-shaped form. Since the body is transparent these animals cannot be detected as they float in search of victims—larvae, copepods and other kinds of zooplankton. Having sighted their prey, the arrow-worms attack

Purple heart urchins, lacking organs for mastication, feed basically on mud.

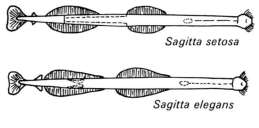
Sagitta setosa

Sagitta elegans

Among the most voracious consumers of plankton are the predatory arrow-worms. Because the various species have marked preferences for different types of water they furnish information to oceanographers about the degree of salinity and the movements of currents. Thus *Sagitta setosa* floats in the colder waters of the North Sea, whereas *Sagitta elegans* is an inhabitant of the Atlantic Ocean.

Among the crustaceans only the lobster is larger than the slipper lobster of the Caribbean (*above*). Another huge inhabitant of shallow waters is the fan shell (*below*), a bivalve mollusc which may grow to 3 feet long.

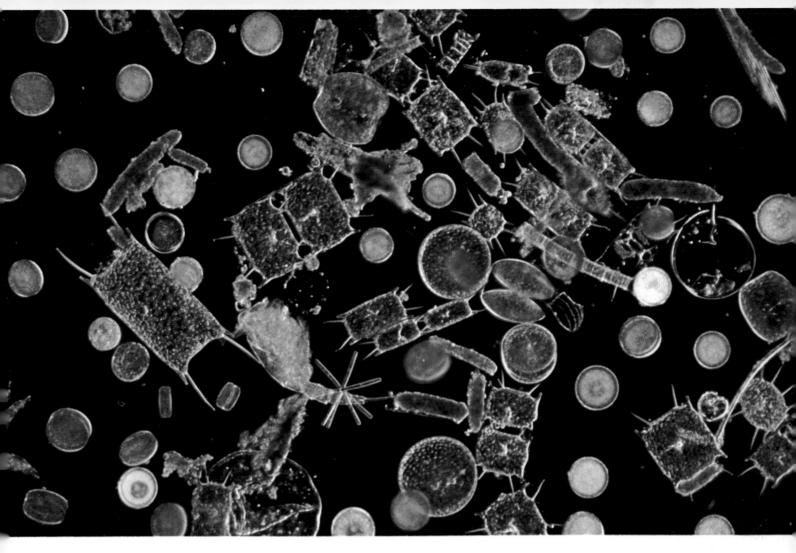

and swallow them in a flash. Other animals creating havoc in these communities are jellyfishes and comb jellies.

Other constituents of zooplankton are the larvae of animals which, when they become adult, will live either on the sea bed or in the open sea, according to their identity. Because of their unusual biology they merit brief mention here. The parents of these larvae will have released extraordinarily large numbers of eggs, sometimes totalling millions, but relatively few of these will have hatched, for several reasons. The yolk sac of these eggs will often fail to provide sufficient nutrition for the proper development of the larva which therefore has to find other forms of food. The comparative scarcity of such food, combined with the number of dangerous predators lurking in the sea, will result in an extremely high death rate among such planktonic larvae.

Even if they manage to survive the larval stage the troubles of the adults are not over, for when they head for the ocean bed they are again at the mercy of predators, different from those previously encountered but none the less capable of greatly reducing their numbers. The consequence of these diverse predatory activities is that sometimes more than 90 per cent of the new generation will succumb.

Study of the photic zone shows, therefore, that life and death are delicately poised and that the natural balance of this ecosystem depends upon the working of minute mechanisms which enable organisms to produce energy by feeding on one another. Only by understanding how these complex marine food chains operate can man plan his fisheries in a rational manner so that they do not cause lasting damage to the inhabitants of the ocean and thus squander a rich heritage.

The abundance of mineral elements and sunlight are the main factors responsible for the development of phytoplankton in the waters of the continental shelf. The greater part of this microscopic plant matter is consumed by zooplankton, tiny marine animals, and thus forms the basis of all the complex food chains of the ocean.

CHAPTER 9

The fishes of the shallow seas

For reasons already explained, the waters of the continental shelf constitute an ecosystem individual enough to merit separate study. When we come to look at the fishes of these areas it is immediately obvious that they can be divided broadly into two categories–those that swim near the surface or at varying depths below, and those that live on or close to the sea bed. These options have led the two groups concerned to acquire very different adaptive mechanisms, superficially evident in their body structure. But whereas the fishes of the benthic zone, never straying far from the sea floor, display marked features of specialisation and adaptation, those of the pelagic zone, ranging far and wide over the sea, are more difficult to classify individually. Because many of them are closely and frequently dependent on the substratum, it is not always easy to decide which of the two groups they belong to; and this ambiguity is especially characteristic of the fishes inhabiting the shallow waters of the continental shelf.

On the high seas, where species are clearly differentiated, no such confusion can arise. Many thousands of feet separate the surface from the ocean bed and since no light penetrates the depths, the inhabitants of surface waters and the abyss are complete strangers to one another. On the continental shelf, however, sunlight reaches down almost to the lowest levels and the food chains depend not only on floating phytoplankton but also on many-celled plants and algae growing on the bottom. The result is that individuals representing both communities may mingle freely with one another, especially near the coasts. This intermingling operates horizontally as well as vertically, so that there is no rigid demarcation line between the pelagic environment of the coasts and the high seas. Fishes may

Facing page: Many pelagic fishes of the continental shelf form enormous shoals. Typical species are herrings, important food fishes. The gregarious tendencies of such species have been the object of much study by ethologists, but whether they are designed for self-defence or for saving energy has not been definitely established.

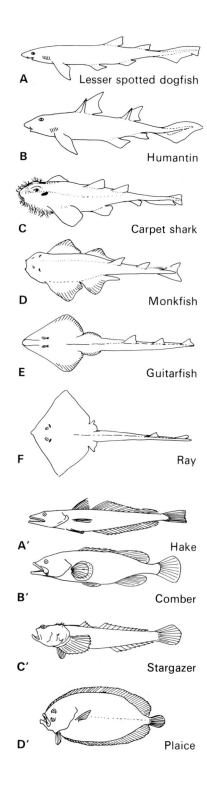

Both cartilaginous and bony fishes have adapted to life in the benthic zone by a progressive flattening of the body. In the former group (*above*) this flattening is from top to bottom, most strikingly evident in the rays; in the latter group (*below*) it is from side to side, reaching its climax in plaice and related species.

roam to and fro at varying distances from the shore so that it is virtually impossible to pinpoint a definite habitat. But by and large the pelagic zone may be said to possess a fairly simple ecosystem, with relatively few species but a large number of short-lived individuals which grow rapidly and are exceptionally prolific. There are of course seasonal fluctuations and all these features are clearly in evidence in cold and temperate waters, less so in warmer seas.

The pelagic fishes of the continental shelf show little specialisation and their principal characteristic is that they are excellent swimmers. Most of them have a simple behaviour pattern which basically consists of moving about more rapidly when attacking prey or fleeing predators. They can be classified in two groups—those that are gregarious and form shoals (although such forms of organisation are primitive and not to be interpreted as being genuinely social) and those that remain solitary, these generally being powerful predators that live at the expense of the former species.

The benthic zone, by contrast, offers a multitude of ecological niches. Here, although there are many more species, the number of individuals per species is fewer. These fishes are more diversified and specialised than those of the pelagic zone, exhibiting an enormous range of forms and life patterns. Various groups, although unrelated to one another, have adapted to the problems of existence at this level in highly ingenious, yet very different ways. The manner in which they have, as it were, conquered and settled the bed of the sea is perhaps the most fascinating aspect of life on the continental shelf.

The sea bed, broadly speaking, is of two principal types, rocky or coralline on the one hand, sandy or muddy on the other. The animals of the former environment are far more complex, suggesting that the evolutionary processes have been allowed to develop more freely and richly. It is evident that species so singular in shape or colour (although these have a precise function) could only have evolved in surroundings such as these, where the terrain is uneven and extremely varied in structure. By contrast, the sandy or muddy bottoms that cover the major areas of the continental shelf, comparable in their scenic monotony to the steppes, harbour a much more uniform fish community. There are really only two types of fish in this environment, those with a fusiform body capable of swimming rapidly in pursuit of prey or to avoid enemies, and those with a flattened body which survive by remaining still, undetected by reason of camouflage. The range of colours is very small. The former are greyish or silvery, the latter brownish or yellowish so as to blend with the ocean floor, or capable of colour-changes. In both groups the back is dark and the belly pale. The fishes of sandy and muddy sea beds are, however, more fertile than those of rocky and coralline regions, so that enormous numbers of individuals of a single species are often present, as is the case, for example, with the cod of cold and temperate seas.

It must be remembered, nevertheless, when describing the fishes of the continental shelf, that the above-mentioned zones overlap and that there are no clear frontiers in this world.

Angler fish
(*Lophius piscatorius*)

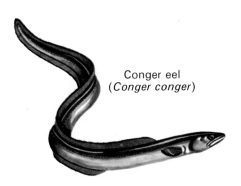

Conger eel
(*Conger conger*)

Moray eel
(*Muraena helena*)

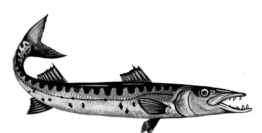

Great barracuda
(*Sphyraena barracuda*)

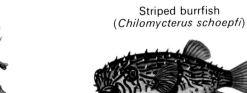

Striped burrfish
(*Chilomycterus schoepfi*)

Sea horse
(*Hippocampus guttulatus*)

Atlantic herring
(*Clupea harengus*)

Hake
(*Merluccius merluccius*)

Boxfish
(*Ostracion lentiginosm*)

Grouper or comber
(*Serranus guaza*)

Bluefish
(*Pomatomus saltatrix*)

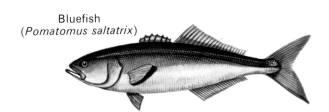

Cuckoo wrasse
(*Labrus mixtus*)

Stargazer
(*Uranoscopus scaber*)

Sole
(*Solea solea*)

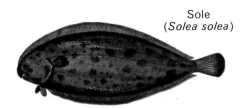

Mackerel
(*Scomber scombrus*)

Red mullet
(*Mullus surmuletus*)

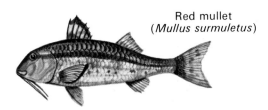

Mysteries of shoal formation

The weak point in the argument that fishes form shoals in order to protect themselves from enemies is that the same gregarious instincts are found among certain predators. This picture shows a group of Californian barracudas, similar shoals being formed by related species. It lends support to the argument that fishes in groups expend less energy than those leading a solitary life.

Many coastal pelagic species collect in vast shoals that are often made up of millions of fishes. This gregarious behaviour is of course a tremendous boon for commercial fishing which concentrates on those few familiar species that are most abundant. Zoologists do not know, however, precisely what motivates the formation of such shoals and to what extent individual fishes are thereby protected. There are theories that the habit is a form of adaptation, enabling such species to withstand the attacks of predators more successfully. Yet it is hard to verify this hypothesis since, if anything, the presence of a shoal might be expected to make an attacker's task easier; and it fails to explain why certain predatory species such as barracudas themselves form shoals. Other zoologists suggest that it may have some connection with reproduction, but this does not allow for the existence of single-sexed groups. Can it perhaps facilitate the job of finding food? This is all very well for fishes on the edges of the shoal but not for those in the centre which have practically no freedom of independent movement. There is a further theory which has not so far been tested experimentally but which seems plausible, namely that swimming in shoals may be dictated by hydrodynamic necessity, consuming less energy than on an individual level.

Apart from motivation, another interesting problem relating

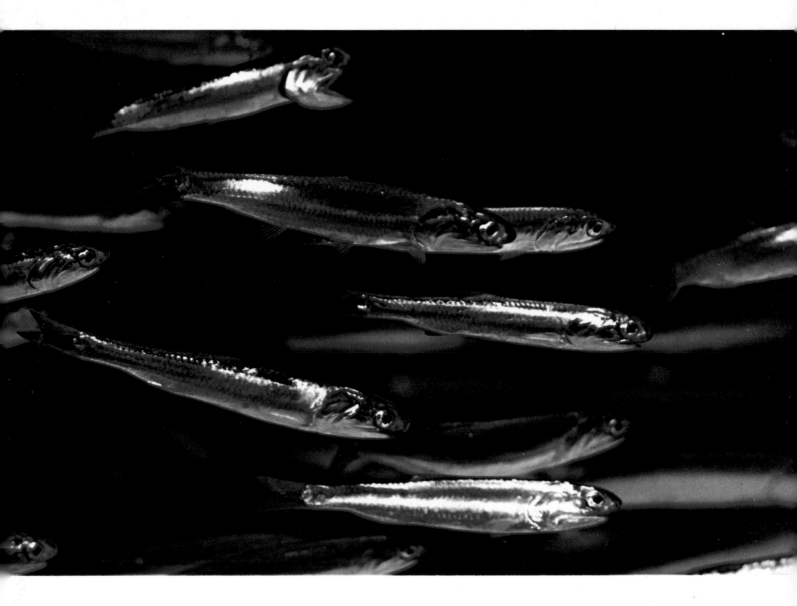

to shoals is that of coordination. None of the theories yet advanced can fully account for the amazing spectacle of many thousands of fishes of the same species all moving together in one direction as if driven by a common impulse. What has been established, however, is that each individual follows its companions, guided principally by vision and the sensing of changing water pressures and movements, thanks to the mucous organ along its flanks known as the lateral line. The result is that each fish keeps a certain distance from the next one by some interplay of attraction and repulsion. For those subscribing to the hydrodynamic theory this coincides with the distance at which expenditure of energy would be minimal.

Whatever the true motivation or function of shoal organisation, it is obvious that it is no matter of chance, as has been demonstrated experimentally by Evelyn Shaw with groups of fry. Her conclusion was that the gregarious tendencies of an individual increase gradually as it grows. Thus when the young fish measured about one-quarter of an inch its immediate reaction, faced with another fish of the same species, no matter what the circumstances, was to flee. At about three-eighths of an inch it would let a congener approach to within an inch or so of its tail, swimming along with it for a couple of seconds; but should it catch a glimpse of the other's head, either from the side or the front, the instinct for flight persisted. Soon afterwards, however, it would swim head to tail for five or

Anchovies, like other members of the herring family, habitually swim in shoals and are fished in enormous quantities. Most of their food consists of plankton, filtered by the gill rakers.

Geographical distribution of the Atlantic herring.

ATLANTIC HERRING
(*Clupea harengus*)

Class: Osteichthyes
Order: Clupeiformes
Family: Clupeidae
Length: up to 16 inches (40 cm)
Diet: copepods, pteropods, larvae of crustaceans and other planktonic organisms
Number of eggs: up to 47,000
Longevity: 20–25 years

Slender body. General colour silvery-grey. No lateral line. Ventral fin situated a little behind front edge of dorsal fin. Gill cover lacks radiating lines.

---

ten seconds with a companion. At just under half an inch the alevin might itself approach the tip of another's tail, both vibrating their bodies. This indicated a readiness to swim in line or beside each other for up to a minute. Sometimes they would be joined by a third or fourth fish; and having attained a length of half an inch groups of some ten individuals would form. At this point the distance between each fish ranged from three-eighths of an inch to one and a half inches; but when it had put on another fraction of an inch the intervening space was never more than about five-eighths of an inch.

In the light of this information it seems obvious that shoal formation is a gradual process related to individual growth and awareness, determined, as it were, genetically.

The typical example of fishes accustomed to moving about in shoals is furnished by the members of the family Clupeidae, among which are herrings and sardines.

Migrating shoals

Fishermen since ancient times have been aware that shoals of Atlantic herrings (*Clupea harengus*), migrate annually from north to south. Now we have a different picture. Herrings flourish in water temperatures of 6-15°C (43-59°F). Each year the Gulf Stream moves north-east across the Atlantic, reaching successively in summer the coasts of France, the British Isles, the Low Countries, Scandinavia and Iceland, warming the surface waters. Herrings live in colder waters, so they desert the surface waters for deeper levels. When, in summer and autumn, the warm waters withdraw, the shoals appear in the surface waters first off Shetland, then in successive areas in the North Sea and finally off the coast of Brittany in January, giving an impression of a southward migration.

The various races of herring, identifiable by the number of vertebrae, never breed with one another even though they may occupy the same spawning ground, for their spawning periods differ. The reproductive system is simple. Groups of fishes of both sexes come together in suitable areas and release their sex-cells simultaneously, this split-second timing evidently being stimulated by special fin movements. The fertilised eggs are then left to their fate.

Each female lays between 20,000 and 47,000 eggs, each of them about one-sixteenth of an inch in diameter—a small number in comparison with other species which lay eggs by the million. The modest quantity suggests that the eggs are in some measure immune to the attacks of animals that normally consume this type of food. If the water temperature ranges from 11°C (52°F) to 14°C (58°F) the eggs will hatch within eight or nine days; but if it is as low as 0°C (32°F) the interval is about forty-seven days. At birth the larva measures approximately one-eighth of an inch. It is transparent, has neither mouth nor gills, and feeds on the remains of the yolk sac. Growth is rapid; within a month and a half it measures almost half an inch and after a year will be more than two inches long.

Herrings feed on plankton, by filtering water through the

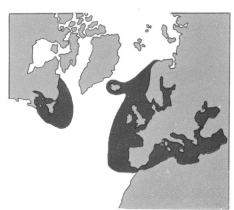

Geographical distribution of mackerel.

Very sensitive to water temperature, most members of the mackerel family make seasonal migrations.

bony gill arches. The inner sides of the arches (as against the outer edges which bear the gills) carry a series of comb-like projections, the gill-rakers. These sieve any floating particles, including plankton. Larger plankton is, however, deliberately seized with the jaws.

The common mackerel (*Scomber scombrus*) belongs to the same family as the tuna or tunny fish. Like the latter it lives in shoals and undertakes annual migrations. The majority of mackerel spend the winter at great depths in the North Sea and neighbouring waters, dispensing with food during this period and recovering their appetite only in spring. Zooplankton – the basic element of their diet – is filtered from the sea water in the same manner as that of herrings, by means of the gill-rakers. At this season they swim nearer the surface and when they reach coastal waters where the temperature is between 11 °C and 14 °C they commence spawning. In the Mediterranean this will occur in March and April; off the south coast of England, the north coast of France and in the North Sea it will continue into June. Each female lays about half a million eggs (up to one-sixteenth of an inch diameter) in shallow water, and these hatch within about six days. The newly hatched larvae measure some one and three quarter inches, the fry remaining close to shore until autumn. Within two years they attain a length of about eight inches.

When the spawning season is over the adult mackerel become predators, roaming the seas in small groups, feeding largely on herrings and other medium-sized species of fishes. In August or September they once more leave coastal waters and prepare for a new period of winter abstinence.

Mackerel occupy a key position in the food chains of larger coastal pelagic fishes. By their large-scale consumption of zooplankton and small fishes they transform this apparently insignificant food into a large mass of protein which sustains superpredators such as tunnies and sharks.

COMMON MACKEREL
(*Scomber scombrus*)

Class: Osteichthyes
Order: Perciformes
Family: Scombridae
Length: up to 20 inches (50 cm)
Diet: small fishes of Gadidae and Clupeidae families, copepods and other constituents of zooplankton
Number of eggs: up to about 500,000

Fusiform body. Colour silvery-green; black marks on back. Two widely separated dorsal fins, with 4–6 finlets behind second dorsal and anal fin.

The larger Spanish mackerel (*Scomber colias*) frequently shares the same feeding and spawning grounds as the related common mackerel. In general, however, it has a rather wider distribution, with a preference for warmer waters. The ecological roles and biology of the two species are similar.

The ferocious bluefish

Immediately above the mackerel in the food chain are a number of fishes, of which the bluefish (*Pomatomus saltatrix*) may be selected as a representative species. Despite their modest size (they measure at most 32 inches) bluefish are fierce and highly ingenious predators. Hunting in groups, they create havoc in coastal waters, attacking any fishes smaller than themselves, either swallowing them whole or slicing them into bits beforehand. So voracious are they that there is not always time to devour all they kill, and a shoal will often leave a trail of bloody fragments in its wake.

Some really astonishing figures have been quoted to suggest the scope of bluefishes' predatory activities. On the reckoning that a thousand million of these hunters arrive in the western Atlantic in the summer season, and assuming that each gets through ten fishes a day, the total number of victims in a single season will be in the region of one million two thousand million! Even if these figures are somewhat exaggerated, there can be no doubt of the significance of the role of the bluefish as predator wherever it roams.

Very little is known about the migratory habits of this species, so that it has not so far been possible to obtain a clear picture of its life-cycle. The only information available concerning reproductive behaviour has been the result of laboratory observations. These indicate that spawning occurs between June and August and that the eggs take from forty-four to forty-six hours to hatch, the larvae measuring only about one-eightieth of an inch. Although at this stage they possess jaws and gills, these are non-functional; furthermore the pectoral fins are absent. Growth, however, is rapid. Within a year the bluefish measures about 20 inches and weighs $1\frac{3}{4}$ lb; by the age of three years the size has increased to 28 inches and the weight to about $3\frac{1}{2}$ lb.

Because of the lack of reliable information concerning their biology, it has not been possible to say to what extent, if any, bluefishes are preyed upon by other species.

The large pelagic predators

The food chains to be found in the open waters of the continental shelf are extremely complex and frequently overlap. But in this environment, as in all others, whether on land or at sea, the dominant roles are played by super-predators. Among the bony fishes the most typical super-predators here are the various barracudas of the genus *Sphyraena*, fierce and powerful carnivores the largest of which may measure up to 10 feet long.

The body of the barracuda, supported by strong muscles, is

SPANISH MACKEREL
(*Scomber colias*)

Class: Osteichthyes
Order: Perciformes
Family: Scombridae
Length: 14 inches (35 cm)
Weight: $4\frac{1}{4}$ lb (2 kg)
Diet: similar to that of common mackerel

Resembles common mackerel but larger eyes and pectoral scales. Blue-grey marks below lateral line.

BLUEFISH
(*Pomatomus saltatrix*)

Class: Osteichthyes
Order: Perciformes
Family: Pomatomidae
Length: up to 32 inches (80 cm)
Diet: all kinds of small and medium-sized fishes

Strong, fusiform body. Back and upper parts of flanks vivid blue-green. Belly silvery-white. Two dorsal fins, the first small with faint rays, the second large and sharply angled, like the anal fin; caudal fin powerful and deeply notched.

BARRACUDA
(genus *Sphyraena*)

Class: Osteichthyes
Order: Mugiliformes
Family: Sphyraenidae
Length: up to about 120 inches (300 cm)
Diet: plankton-eating fishes

Long, cylindrical, powerful body. General colour silvery. Large head with pointed snout; protruding lower jaw. Very strong caudal fin; the single dorsal fin is small and situated well to rear of body, on same level as anal fin.

beautifully streamlined, the head slender, the mouth equipped with rows of exceptionally sharp teeth, the lower jaw protruding beyond the upper one. Barracudas are especially prevalent around coral reefs, roaming in large shoals and provoking immediate panic in the ranks of the other fishes as soon as they put in an appearance.

Little information has been accumulated concerning the biology and behaviour of barracudas. Nothing is known, for example, of their breeding habits although it is assumed that they spawn in the open sea. What reports there are of their behaviour are confined to confrontations with humans, and even these tend to be so sketchy and sometimes contradictory that they cannot necessarily be regarded as typical. It has not been explained why these predators appear to be dangerous in certain waters yet quite inoffensive in others. What is more, it seems that an underwater swimmer may be in less danger if surrounded by a shoal of curious barracudas than if he comes up against a single individual. It has been established that attacks, when they do occur, are not directed against the person but unleashed by the sight of some glittering object attached to the swimmer, such as the chromium-plated rim of his diving mask, his oxygen cylinders, the belt of lead weights used as ballast or even the scales of fishes that he may have slung around his waist after catching them.

The coastal pelagic super-predators also include a number of cartilaginous fishes, notably the sharks. Thus the strange hammerhead sharks (*Sphyrna zygaena*) frequent warm waters

One of the most powerful and greatly feared predators of the continental shelf is the barracuda, the largest species of which measures up to ten feet.

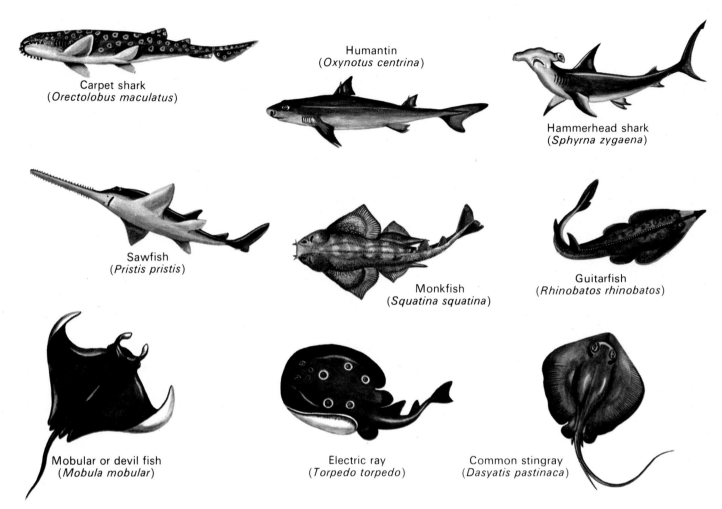

Carpet shark (*Orectolobus maculatus*)
Humantin (*Oxynotus centrina*)
Hammerhead shark (*Sphyrna zygaena*)
Sawfish (*Pristis pristis*)
Monkfish (*Squatina squatina*)
Guitarfish (*Rhinobatos rhinobatos*)
Mobular or devil fish (*Mobula mobular*)
Electric ray (*Torpedo torpedo*)
Common stingray (*Dasyatis pastinaca*)

HAMMERHEAD SHARK
(*Sphyrna zygaena*)

Class: Chondrichthyes
Order: Lamniformes
Family: Sphyrnidae
Length: 120–180 inches (300–450 cm)
Diet: fishes and invertebrates

Back slate-grey or greyish-brown. Head shaped like hammer. First dorsal fin large, close to pectoral fins; second dorsal fin and anal fin small. No spiracles (breathing holes).

---

but also make incursions into temperate seas. These huge fishes are powerful carnivores which may measure up to 15 feet long and weigh 1,750 lb. Like barracudas, they tend to swim in groups of about half a dozen individuals, although in some regions they may also be found alone or in much larger shoals. They have a wide-ranging choice of prey, but whereas it is true to say that there are few species at any depth which are immune, the favourites seem to be mackerel.

Hammerheads are just as unpredictable as barracudas when confronted by underwater divers, attacking them in some areas, ignoring them in others. The difference in attitude may stem, to some extent, from the size of prey to which they are normally accustomed. Thus in the region of the Galapagos Islands, where their victims frequently include seals, the sharks have been known to attack humans more often than in other places where they have adjusted to smaller forms of prey.

The common name of the species was dictated by the very unusual appearance of the head. Several theories have been advanced to explain the probable function of this hammer-shaped structure, one that it may serve as a kind of rudder in the ocean depths, rather like that of a submarine. Another, put forward in 1969 by D. R. Nelson (but which still lacks proof) is that the head is a form of 'stereoscopic' scenting device, providing a wide gap between the two halves of the olfactory apparatus, thus enabling the shark to use its sense of smell most efficiently to locate its victims.

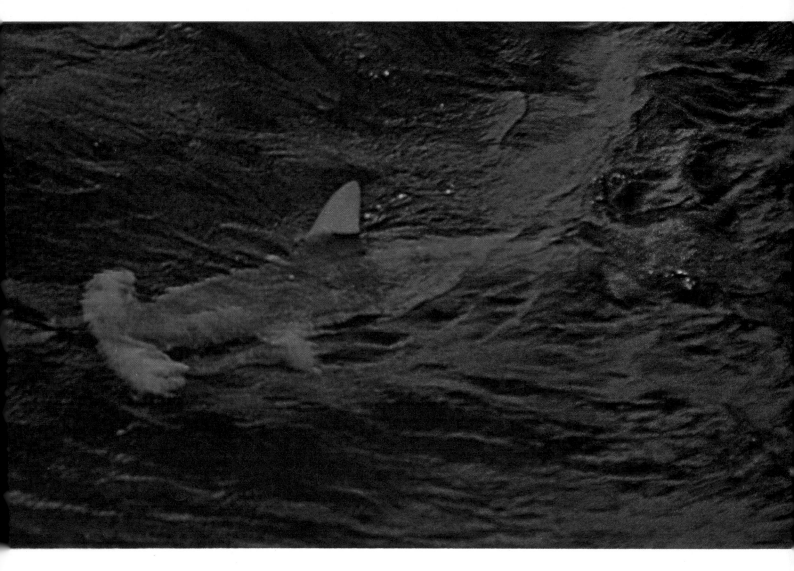

Life on the sea bed

From the viewpoint of zoologists specialising in problems of evolution and animal behaviour, the most interesting fishes of the continental shelf are undoubtedly those that have chosen to live in the shallow waters of the benthic zone. These species display a variety of inventive, often surprising forms of adaptation to their surroundings.

The adjustment to life on the sea bed began almost as soon as the sea was first populated by fishes which, according to fossil remains, originated in fresh water. Among the most primitive groups are species which already show signs of having adapted to benthic conditions and of being greatly influenced by this environment. It is certainly evident among the sturgeons of the family Acipenseridae, considered to be the oldest of living fishes. The surroundings in which they evolved account for a number of interesting modifications, one of the most astonishing relating to the sense of taste. For whereas the majority of vertebrates have a tongue for this purpose, sturgeons possess taste organs in the shape of several barbels growing in front of the mouth, an extremely useful position in the case of a species which seeks its food on the bottom. It is significant too that another quite unrelated group, that of the gurnards (family Triglidae) have evolved in convergent fashion by developing a sense of taste outside the mouth. In their case the taste buds are attached to the modified rays of the pectoral

The hammerhead shark is immediately recognisable by the very unusual shape of its head. There is much discussion among zoologists as to whether it serves as a kind of rudder or whether it may be an aid to scenting prey.

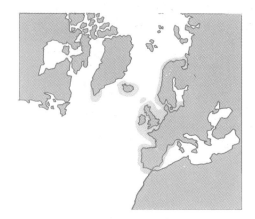

Geographical distribution of plaice.

PLAICE
(*Pleuronectes platessa*)

Class: Osteichthyes
Order: Pleuronectiformes
Family: Pleuronectidae
Maximum length: 38 inches (95 cm)
Weight: 15¼ lb (7 kg)
Diet: benthic invertebrates
Number of eggs: 50,000–500,000
Longevity: about 50 years

Flat, smooth body rounded in outline, covered with small scales. Upper part brown with red spots. Crest of bony projections on head, extending between eyes. Rounded caudal fin.

HALIBUT
(*Hippoglossus hippoglossus*)

Class: Osteichthyes
Order: Pleuronectiformes
Family: Pleuronectidae
Maximum length: 120–160 inches
(300–400 cm)
Weight: maximum 660 lb (300 kg)
Diet: fishes, large crustaceans, cephalopods
Number of eggs: 2,000,000–3,500,000
Longevity: 40–50 years

Flat, elongated body. Lateral line strongly curved at front. Eyes situated on right side. Dorsal fin begins at level of upper eye. Caudal fin slightly notched.

Facing page: Two very different inhabitants of the shallow waters of the continental shelf are the lesser spotted dogfish (*above*) and the flounder (*below*).

fins. These are shaped like fingers and enable the fishes virtually to 'walk' along the bottom. Because of this permanent contact with the substratum gurnards can be constantly aware of the presence of food without interrupting their normal run of activities.

By studying such modifications as they appear in various surviving species it has been possible for zoologists to discover how, by slow stages, the principal groups of benthic fishes adapted to such conditions. The different phases of the process can best be seen in the body shapes of the fishes involved. One group comprises a large number of species which have been transformed very little, having retained the typical fusiform shape (rounded in the middle, tapering towards either end) so that the changes which have overtaken them have affected their behaviour rather than their outward form. Typical examples are the Gadidae, including cod, haddock and hake. All these fishes feed on the bottom and their ability to hunt and flee predators depends entirely on their swimming powers. Slightly different, and not such effective swimmers, are those species that are more massively built and which seek refuge in rock cavities on the ocean floor, as is the case with the Serranidae, including bass, grouper and comber.

Other branches have been transformed in a much more obvious and spectacular manner. The best way to remain undetected on the sea bed is to have a body which as far as possible blends with the surroundings. Consequently a number of species have become completely flat. The phenomenon has occurred in two groups. The first comprises bony fishes (Pleuronectiformes) such as soles, plaice, flounder and other related species, in which the body is flattened laterally. The second group consists of cartilaginous fishes such as rays and skate, where the same effect has been achieved by a vertical flattening of the body.

There are other groups, perhaps less important, which have also acquired singular features that invariably suit their particular life styles. There is, for example, the sea horse, whose prehensile tail is of enormous value for obtaining food in the fields of seaweed where it lives. The anglerfishes and related species, for their part, are masters of camouflage; and eels, including the conger and moray, are likewise adjusted in their own fashion to life on the ocean floor.

Beware of sharks

The cartilaginous fishes (Chondrichthyes), despite their long history, still occupy a vitally important place in the overall pattern of marine life. They have produced a large variety of lines that are well adjusted to conditions in the benthic zone. The different processes which have led such fishes to harmonise so perfectly with their environment can be followed through successive phases. Indeed as the group continued to evolve, the ecological niches occupied by their more primitive forerunners have remained the same, so they have become better adapted than other species. As a result, specialists have been able to retrace the successive stages in their evolution.

Geographical distribution of the lesser spotted dogfish.

LESSER SPOTTED DOGFISH
(*Scyliorhinus caniculus*)

Class: Chondrichthyes
Order: Lamniformes
Family: Scyliorhinidae
Length: 24–32 inches (60–80 cm), maximum 40 inches (100 cm)
Diet: fishes, crustaceans and molluscs
Number of eggs: 18–20

Very slender, long body. General colour pale chestnut, almost white on belly. Back completely covered with small brown spots. Rear tips of ventral fins long and pointed.

HUMANTIN
(*Oxynotus centrina*)

Class: Chondrichthyes
Order: Lamniformes
Family: Squalidae
Length: up to 60 inches (150 cm)
Diet: benthic animals

General colour greyish-brown. Prominent nostrils on either side of mouth, which is always open. Heavy head. Five gill openings on either side. Two high dorsal fins, each with a strong spine. Large pectoral fins. One anal fin.

The first stage in this long evolutionary journey is exemplified by what are the best known and certainly the most feared of all cartilaginous fishes—the sharks. The Carcharinidae are one of the many shark families represented among benthic communities and some of these animals are even permanent freshwater inhabitants. Probably the most typical is the Lake Nicaragua shark (*Carcharinus nicaraguensis*) which lives in the South American lake of that name. But among the sharks of coastal shallow waters the most characteristic are without doubt those belonging to the family Scyliorhinidae, commonly known as dogfishes. These are really miniature sharks which seldom measure as much as 3 feet and are to be found principally at depths of between 30 and 250 feet in regions where the sea bed is sandy and covered with algae. The diet of these species is extremely varied and not much of it consists of fish. The main food items are worms, sea cucumbers, octopuses and a number of other benthic animals. Inactive during the day, dogfishes hunt at night, and their common name probably derives from the fact that, like dogs, they are guided to their prey by scent. The breeding season is a long one, lasting from November until July. Fertilisation is accomplished internally by means of copulatory organs characteristic of sharks, namely stiff, modified portions of the pelvic fins. Births are staggered throughout the year but are most numerous in spring and summer.

A second stage in adaptation to life in shallow waters is represented by the bull-headed or horn sharks of the family Heterodontidae, recognisable by their heavy body, the protuberances that appear above the eyes and the reputedly venomous spine which juts up in front of each of the dorsal fins. This last feature is in itself proof of their specialisation. In fact, observation of fishes living in these shallow seas confirms the belief that in a general way such venomous spines are used in self defence. It is thanks to their presence that scientists have been able to identify certain fossils and to conclude that this group has remained virtually unchanged for some 150 million years, a state of stability which bears witness to its perfect adaptation.

The diet of these bull-headed sharks is scarcely known, but is probably made up of a wide range of benthic animals, among which must surely be included certain hard-shelled molluscs, for some of the sharks have teeth that are flattened and which fit close together like stones in a mosaic, or pavement, and which form the perfect instrument for crushing hard shells.

The mating of these sharks is preceded by a rough form of courtship in which the male bites the female in various parts of the body, then curls himself around her and fertilises her with his copulatory organs. The eggs are laid one after another at intervals of two months, the singular feature of them being that they are cylindrical capsules, metallic chestnut in colour, with a curious spiralled edge that resembles a helix or the head of a drill. This unusual structure apparently makes it possible for each egg to become buried in the sand or mud, thereby concealing it from predators. After seven or eight months these strange eggs hatch and the young sharks emerge, measuring about 8 inches.

Of all the members of the family, it is the carpet sharks (*Orectolobus maculatus*), closely related to the nurse sharks, which have undergone the most extensive modifications for adapting to life in the benthic zone. The transformation has been such as to give the sharks some resemblance to the anglerfishes but this is simply an example of convergent evolution, for the two groups are unrelated.

The common name of these sharks is derived from the lively pattern of colours all over the body. These markings are, of course, no mere embellishments but serve for purposes of camouflage, blending in a remarkable manner with the sand and rocks of the ocean bed. Indeed the whole structure of this species is most untypical of a shark, for the head and front part of the body are flat and rounded, as are the pectoral fins, so much so that there is a striking similarity to the more typical flat fishes. On the other hand, the tail is thin and the caudal fin very small. Breaking the smooth outline of the front portion of the body are a number of fleshy, branching flaps of skin, resembling algae, jutting out around and behind the mouth. The eyes too are protected and almost completely concealed by folds of skin.

Another curious feature is that these sharks breathe in a manner which is characteristic of the flat cartilaginous fishes such as rays. Behind the eyes are spiracles (breathing holes) and the gill slits are situated at the base of the pectoral fins. The sharks take in water through the spiracles rather than through the mouth, so that there is no risk of their taking in unwanted foreign matter from the ocean floor.

Carpet sharks spend most of their time motionless on the bottom, so effectively hidden that unsuspecting fishes and crustaceans are easily caught.

Little is known of breeding habits, apart from the fact that the sharks are ovoviviparous and very prolific. Reproduction involves an act of copulation and fertilisation is internal.

Between sharks and rays

If certain sharks show signs of adaptation to life in the benthic zone, the rays and related species (Rajiformes) display the completed process, as we shall shortly see. It is not always possible to trace the various stages of evolution of an animal to a given habitat but in this case a valuable clue is provided by two strange species occupying a place somewhere between the principal groups of sharks and rays.

The first is the monkfish (*Squatina squatina*), also known as the angel shark. It has gill openings along the flanks, like sharks, but differs from the latter in possessing a flattened body and a respiratory system characteristic of rays and similar species, water entering through spiracles situated behind the eyes. The front of the body is broad and flat, and since the pectoral and pelvic fins are large and the tail very narrow, the ray-like appearance is even more accentuated.

The monkfish's behaviour is also akin to that of the Rajiformes, with food consisting principally of flatfishes, molluscs, crustaceans and worms; but it emulates the sharks in supple-

Geographical distribution of monkfish.

MONKFISH
(*Squatina squatina*)

Class: Chondrichthyes
Order: Lamniformes
Family: Squatinidae
Length: 35½–47½ inches (90–120 cm),
  maximum 98 inches (250 cm)
Weight: up to 176 lb (80 kg)
Diet: small benthic fishes, molluscs and
  crustaceans
Number of young: 10–25

Colour greyish. Body long and flattish. Large pectoral and pelvic fins. Two dorsal fins and anal fin are small. Spiracles open behind eyes.

CARPET SHARK
(*Orectolobus maculatus*)

Class: Chondrichthyes
Oredr: Lamniformes
Family: Orectolobidae
Length: up to 138 inches (350 cm)
Diet: fishes and crustaceans

Unique appearance for shark, with head and front of body very broad and flat. Front edge of head covered with branching protuberances. Pectoral fins large and rounded; dorsal fins fairly small. Eyes small, protected under folds of skin.

Geographical distribution of electric marbled ray.

TORPEDO or ELECTRIC MARBLED RAY
(*Torpedo marmorata*)

Class: Chondrichthyes
Order: Torpediniformes
Family: Torpedinidae
Length: up to 24 inches (61 cm)
Diet: small fishes, molluscs and crustaceans

Back chestnut, marbled; abdomen white. Body circular in shape. Tail short and thicker than that of other rays. Electric organs on either side of head, visible under skin.

GUITARFISH
(*Rhinobatos rhinobatos*)

Class: Chondrichthyes
Order: Rajiformes
Family: Rhinobatidae
Length: up to 72 inches (183 cm)
Diet: small fishes, molluscs, crabs and other benthic animals

Back grey or brown; abdomen pearl-grey. Front of body flattened, almost triangular in shape, similar to that of rays. Caudal region long, slender and cylindrical, like that of sharks. Two dorsal fins and long denticles down median line of back. Spiracles behind eyes.

menting its diet with pelagic fishes and sea birds. Its movements through the water are also shark-like—effected by sideways flicks of the tail rather than flappings of the pectoral fins. It generally frequents fairly shallow seas, although some species live at depths of up to 4,200 feet. In June or July, however, at the onset of the breeding season, these individuals also make their way back to shallower waters. Females are ovoviviparous and give birth to nine to sixteen young according to their size. Some reports give up to twenty-five.

The second intermediate species between sharks and rays is the guitarfish (*Rhinobatos rhinobatos*), which represents yet another forward step in the progressive adaptation to benthic life. Here too the front of the body is flattened but the pectoral fins surround the head and join in front to form a kind of snout. Mouth and gill slits are located on the ventral side and the spiracles situated behind the eyes. The tail fin is thinner than that of the monkfish but not as narrow and pointed as that of the true rays. The guitar fish, like the monkfish, propels itself with its tail and not with its 'wings'. It too is ovoviviparous, the eggs being hatched within the body and the young born alive.

Rays—electric and non-electric

The cartilaginous fishes best adapted to conditions of life on the ocean bed are the various species with flat bodies belonging to the orders Hypotremata and Rajiformes.

The Hypotremata, made up of the single family Torpedinidae, are commonly known as electric rays, differing in appearance from other rays in that the head, trunk and pectoral fins are linked to form an almost circular disc, the tail being relatively short and stubby. The principal distinctive feature, however, is the presence of specialised electric organs. Although such organs are not exclusive to Torpedinidae, in them they are so highly developed that they are capable of emitting shocks equivalent to 50-60 volts, and even, in the case of larger individuals, of 200 volts. In contrast to other fishes similarly equipped, the organs in electric rays are situated on either side of the head.

The skin of the fish is smooth and naked, and the resistance of the epithelium covering these organs is less than in other parts of the body; thus the electric discharge can be channelled outwards, greatly increasing its power.

The electric organs serve several functions. Firstly, they are obviously dissuasive weapons to be used in self defence. Secondly, they are equally effective for hunting prey. The ray literally enfolds its disc-like pectoral fins around its victim, stunning or even killing it by electrocution. But the most important function is neither attack or defence, but to act as a sophisticated sonar device, working not by means of a simple echo but by recording changes in the electric field surrounding the ray itself—this field being set up by rhythmic activity of low intensity and voltage. So sensitive is the ray that it can locate extremely small obstacles as well as distinguish between objects of similar size but of differing degrees of electrical conductivity.

The gradual adaptation to conditions in the benthic zone of the continental shelf can be followed by examining three typical species of cartilaginous fishes found in these waters. The humantin (*top*) is modified in various ways for catching invertebrates on the ocean floor. The carpet shark (*centre*), well camouflaged and with strange protuberances around the mouth, lies motionless, awaiting prey. The electric marbled ray (*bottom*) is the best adapted of all, representing the cartilaginous flat fishes, with powerful dissuasive electric organs.

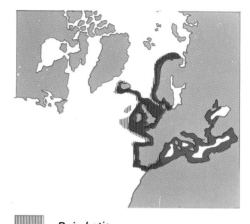

Geographical distribution of the common skate (*Raja batis*) and the thornback ray (*Raja clavata*).

THORNBACK RAY
(*Raja clavata*)

Class: Chondrichthyes
Order: Rajiformes
Family: Rajidae
Length: 27½–47½ inches (70–120 cm)
Diet: crabs, shrimps and small flatfishes
Number of eggs: about 20

Back chestnut, abdomen white. Dorsal region of disc covered with strong spines. Snout forms an obtuse angle.

COMMON SKATE
(*Raja batis*)

Class: Chondrichthyes
Order: Rajiformes
Family: Rajidae
Length: 39½–59 inches (100–150 cm), exceptionally up to 98 inches (250 cm)
Weight: up to 220 lb (100 kg)
Diet: benthic fishes, crustaceans, worms

Back greenish-brown; abdomen dark brown with black streaks and spots. Snout forms an acute angle. Front edge of disc slightly convex. Two small dorsal fins, almost linked, at tip of small caudal region which has two rows of spines.

The technical details of this mechanism are complicated but the end result so incredibly efficient that biologists, engineers and cybernetics experts have devoted time to studying the phenomenon, with a view to developing a similar type of sonar apparatus which can be used for military ends in detecting submarines.

In spite of this fearsome equipment electric rays are not aggressive but quite peaceful animals which usually confine their predatory activities to small crustaceans and other creatures of the sea floor. Most of the time they remain motionless but when they do swim they use the tail for propulsion.

The Rajiformes comprise a number of families of non-electric rays, the best known of which are the rays and skates of the family Rajidae, usually found on sea beds of gravel, sand or mud. In their case swimming is effected only by means of undulating movements of the large pectoral fins, unassisted by the tail. Food consists in the main of small benthic animals and fry, but they sometimes capture larger prey by enveloping them with their 'wings'. Although at certain times of the year they make for deeper waters these journeys are not migrations in the true sense. The thornback ray (*Raja clavata*) ascends to shallower levels in the winter but during the same season the common skate (*Raja batis*) does precisely the opposite.

Reproduction is a much more haphazard affair than for other cartilaginous fishes. Once the eggs are laid they are left to pure chance; but those of skates are large and rectangular in shape, with hook-like tendrils at each corner which catch on to seaweed and anchor the egg-case. The baby looks like a miniature version of the adult, born with fins folded over the back.

The description 'non-electric' is not strictly accurate, for the Rajidae do possess a low-powered electric organ (emitting about 4 volts) in the tail, serving as a locating apparatus rather than as an offensive or defensive weapon.

Many rays are found, often half-buried, in mud or sand; but the eagle rays of the family Myliobatidae normally lead a pelagic existence, moving through the water by flapping their pectoral fins, the effect being that of a heavy, slow-flying bird. They feed mainly on crabs and bivalve molluscs. Because their teeth are flat, arranged somewhat like a mosaic and particularly suitable for grinding purposes, the tough carapaces and shells are no great problem. Other food items include worms and fishes. The very long, slender tail has a venomous spine at the base, sufficiently flexible to be aimed in any direction to ward off attacks but becoming less mobile with increasing age. This spine, however, is always sharp for when it becomes blunted it is replaced by a new one. Many individuals have two, three or even four spines. Since the species are ovoviviparous it might be supposed that the presence of these spines could be inconvenient or dangerous at time of birth; but the process of evolution has found an answer to this problem, for the spine of the newborn eagle ray is very soft and incapable of causing the mother injury. It becomes hard and sharp, however, as soon as it makes contact with the sea water.

The stingrays belonging to the family Dasyatidae are also

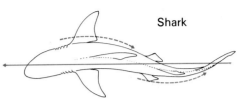

Sharks and rays have different methods of propelling themselves through water, the former by means of lateral movements of the whole body (using the pectoral fins as stabilisers), the latter keeping the body still but undulating their large wing-like pectoral fins. Among eagle rays these fins are flapped somewhat in the manner of birds' wings.

Two typical cartilaginous fishes, their bodies almost completely flat, are the thornback ray (*above*) and the eagle ray (*below*).

Geographical distribution of stingrays.

STINGRAY
(*Dasyatis pastinaca*)

Class: Chondrichthyes
Order: Rajiformes
Family: Dasyatidae
Length: 19½–39½ inches (50–100 cm), maximum 98 inches (250 cm)
Weight: about 22 lb (10 kg)
Diet: fishes, crustaceans, molluscs

Back bluish-black. Tail long, slender; no dorsal fins. Powerful venomous barb halfway down tail.

EAGLE RAY
(*Myliobatis aquila*)

Class: Chondrichthyes
Order: Rajiformes
Family: Myliobatidae
Length: up to 59 inches (150 cm)
Diet: crabs, worms, fishes

Whip-like tail with toothed, venomous spine, in front of which is a small dorsal fin. Large wing-like pectoral fins. Head well detached from rest of body.

MANTA or DEVILFISH
(*Mobula mobular*)

Class: Chondrichthyes
Order: Rajiformes
Family: Mobulidae
Length: 118–236 inches (300–600 cm)
Weight: up to 2,200 lb (1,000 kg)
Diet: plankton

Back dark grey. Whip-like tail. Two horn-like projections in front of mouth.

equipped with a venomous barb in the tail. It too is extremely mobile and can be directed, according to species, forwards, backwards or along the flanks of the fish. The tissue which secretes the venom is located in a gland at the base of the spine and runs along grooved channels to the tip. When the stingray attacks a victim it drives the barb deeply into the wound, injecting the toxic substance and at the same time leaving in it fragments of tissue. Both venom and tissue take some time to be replaced, so that the gravity of the wound will depend on whether the stingray has already recently delivered an attack. The poison works quickly and affects heart, lungs and nerves. Although seldom fatal to humans, it is very painful and may have a temporarily paralysing effect.

Stingrays, like related species, are also ovoviviparous, the eggs hatching in the oviduct. At first the young live on substances contained in the vitelline sac. Later the oviduct walls, with their numerous ducts and vessels, emit a nutritious fluid which the young absorb through the mouth and spiracles. It is a highly evolved feeding mechanism, analogous to the placental system of mammals.

Giant rays

Although the flat cartilaginous fishes are masterpieces of adaptation, they include in their ranks two strange groups of much more original character. Included in the first of these is the manta or devilfish (*Mobula mobular*), one of the giants of the ocean, which may measure up to 20 feet across (from tip to tip of its 'wings') and weigh approximately one ton. Yet despite its enormous size and terrifying appearance, this ray feeds peacefully on plankton, scooping it into its mouth with the aid of two horn-like, flexible projections at the edges of the pectoral fins. Although photographs have been taken of divers clinging perilously to the upper jaw of mantas, the risk is minimal, for these fishes refrain from attacking underwater swimmers and in fact will not even defend themselves. The only danger to the diver is from collision with the huge body.

It is not unusual for a group of these giant rays to leap repeatedly, in most spectacular fashion, clear out of the water, each being followed by a tremendous thudding sound as one heavy body after another crashes down against the surface. The purpose of this strange 'dance' is unknown but many zoologists believe that it may be linked with birth (ovoviviparous), each leap coinciding with the expulsion of a baby from the mother's body.

Although the flat shape of the manta's body might suggest that it too lives on the sea floor, it is a pelagic species; and it is surprising to note that the evolutionary process responsible for producing forms (such as those of smaller rays) ideally suited to benthic conditions, should have been sufficiently flexible to make use of much the same body form for a gigantic pelagic fish. Yet there are contrasts within the family itself. Alongside the devilfish we find the pygmy Australian manta (*Mobula diabolis*) with a 'wingspan' of only about 2 feet.

Mantas are cartilaginous fishes which lead a pelagic existence. In this photograph a group of bony fishes known as remoras are seen clinging to the underside of the manta's body, each attached by a flat suction disc on the top of its head. Remoras are not parasites but are freely transported by rays, sharks and other huge fishes.

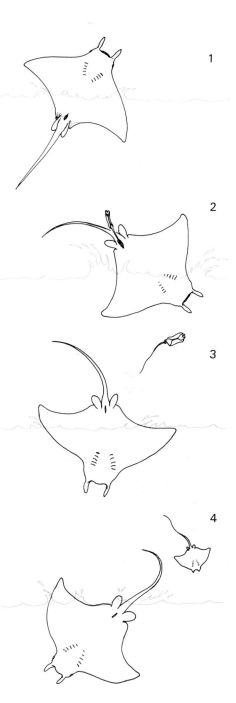

Some zoologists claim that the leaps of mantas out of the sea signal the birth of a baby. Presumably, if this is so, once born the baby opens its 'wings' and flops back into the sea with its mother.

The other curious representatives of the order Rajiformes belong to the family Pristidae. Commonly known as sawfishes, they should not be confused with fishes of the same name which are members of the family Pristiophoridae. The latter have gill openings along the flanks and do not measure more than about 4 feet and their saw-like teeth are alternately short and long. The sawfishes of the family Pristidae, however are considerably larger, bigger even than the mantas, for they measure 20–30 feet in length and may weigh up to 2 tons. The rostrum or snout, with rows of sharp teeth arranged as in a saw on either side, takes up fully one-quarter or even one-third of the total length of the body which is similar in form to that of a shark but a little flatter.

Sawfishes are creatures of the ocean floor, probing with their long snout for food on the muddy bottom. Occasionally they will attack a shoal of fishes, scything their head from one side to another as they cut their victims to pieces. The side-to-side movements of the rostrum appear to be quite instinctive and automatic, apparently triggered off simply by the presence of food. This conclusion is supported by incidents which have occurred in aquaria in which keepers have been injured by these normally peaceful fishes at feeding time.

Sawfishes are ovoviviparous. As is the case with the eagle rays, nature has seen to it that the mother is not injured when her baby is born, for the 'saw' of the latter is not only very soft at this stage but also covered by a membrane which becomes detached soon after birth.

Bony fishes of the benthic zone

The bony fishes, not as primitive as the cartilaginous fishes, have also engendered a multitude of species which have come to occupy different ecological niches in the benthic zone of the continental shelf. They too can be classified in groups corresponding roughly to the relative degree of adaptation—as shown in body shape—to their surroundings.

One group of such fishes is represented by species whose body structure has been influenced little, if at all, by the conditions prevalent on the ocean floor. A good example is the hake (*Merluccius merluccius*). Although its chosen habitat is the sandy or muddy sea bed, it retains the typical fusiform shape, with powerful teeth and long dorsal and anal fins. Unlike fishes that have undergone greater modification, hake move about a good deal, both on a daily and seasonal basis. During the daytime they stay close to the bottom without feeding; but at night they move upwards, swimming at varying depths and paying occasional visits to the surface, in pursuit of prey. In winter and spring the majority of these fishes remain at depths ranging from about 400 to 1,800 feet; but during the spawning season they come up to shallow waters.

The other group of fishes which shows few signs of modification is that of the Labridae or wrasses. These are typical fishes of the coral reefs and rocks, being notable for their brilliant colours. Indeed these gaudy hues pose a problem which has not thus far been resolved by zoologists; for, as they point out, the display of such spectacular livery would seem to be as likely to attract predators as to keep them away. It is possible, however, that the wrasses are sufficiently concealed from enemies by the many cracks and crevices of the rocks and coral, so that the bright colours have been primarily evolved for other purposes, such as the defence of their territories which they do with a fury quite astonishing in fishes of their size. The combination of brilliant colours and the habit of using a crevice in the coral or rock would be consistent with this hypothesis.

It is worth mentioning here certain species which exhibit unusual habits, such as various sand smelts of the family Atherinidae, spawning at high tide and leaving their eggs buried in the sand. Hatching coincides with the next high tide. The sea, as it goes out, takes the alevins with it. The timing is so perfect that neither adults nor fry are ever left high and dry. Other strange fishes are the meagres of the family Sciaenidae which have the odd habit of emitting loud noises (the purpose of which is quite unknown) which are made by vibrating the swim bladders.

Another large and typical family found on rocky sea beds is that of the Serranidae, including combers and near relatives. They generally live in tropical seas but some are found in temperate waters. As a group they are large fishes with a heavy body and sharp teeth. Some are very active, others sedentary, but all are dangerous predators. One surprising feature is that in some species the young are predominantly female, producing normal eggs; but by the time they reach two to five years of age, according

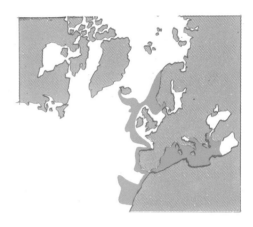

Geographical distribution of hake.

HAKE
(*Merluccius merluccius*)

Class: Osteichthyes
Order: Gadiformes
Family: Merlucciidae
Length: up to about 40 inches (100 cm)
Weight: 22 lb (10 kg)
Diet: herrings, sardines, anchovies, mackerel and other shoaling fishes

Fusiform body. First dorsal fin short, second dorsal fin and anal fin long. Jaw slightly protruding. No barbels. Mouth cavity and gills black.

STARGAZER
(*Uranoscopus scaber*)

Class: Osteichthyes
Order: Perciformes
Family: Uranoscopidae
Length: 10–12 inches (25–30 cm)
Diet: small crustaceans and fishes

Brownish colour. Eyes and mouth situated on top of head; back flattened. First dorsal fin short, second very long, with anal fin slightly shorter; rounded caudal fin; large, rounded pectoral fins. Worm-like filament used for attracting prey can be hidden in mouth. Hard, venomous barb situated on either side of edge of gill cover.

Facing page: A large and varied fish community is to be found among coral reefs. As with so many fishes of the ocean floor, many species are coloured in such a way that they blend perfectly with the surroundings.

The grotesque-looking anglerfish, hunting on the sea bed, uses a long filament, tipped with a flap of skin, to lure unsuspecting prey into its enormous mouth.

to species, all individuals possess rudimentary (non-functional) seminiferous tissues as well as mature female genital organs. Between the age of seven and ten years they are transformed into males despite the fact that they still possess egg-producing tissue which is, in its turn, non-functional. This is only one of many instances of change of sex in fishes that have come to light in recent years.

Zoologists do not know much about these extraordinary sexual changes. In a few species they affect every individual; but although it is generally agreed that signs are evident in all species, the numbers of individuals so transformed are variable. Another fascinating characteristic of wrasses is their ability to change colour very rapidly.

Among the red mullets of the family Mullidae adaptation to ocean floor conditions is more pronounced and these fishes too are able to change colour. Like sturgeons and gurnards they have fleshy barbels in front of the mouth which are sensitive organs of touch and taste.

Flat bodies and shifting eyes

We come now to groups of bony fishes which are even better suited by evolution to live on the sea floor, their bodies having undergone profound transformation in one way or another. In the Uranoscopidae, commonly known as stargazers, the eyes

Facing page: Red mullets, capable of sudden colour changes, possess barbels which are sensitive organs of touch and taste, valuable for locating prey.

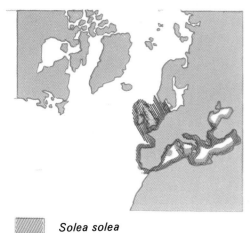

Solea solea

Mullus surmuletus

Geographical distribution of sole (*Solea solea*) and red mullet (*Mullus surmuletus*).

SOLE
(*Solea solea*)

Class: Osteichthyes
Order: Pleuronectiformes
Family: Soleidae
Length: 12–16 inches (30–40 cm)
Weight: about 11 ounces (320 g), maximum 6½ lb (3 kg)
Diet: soft-shelled lamellibranches, bristle worms, crustaceans and some small fishes
Number of eggs: 100,000–150,000
Longevity: more than 20 years

Upper part brown, lower part white. Well developed pectoral fins. Small nostrils, set wide apart, on blind side. Very rounded snout and mouth.

RED MULLET
(*Mullus surmuletus*)

Class: Osteichthyes
Order: Perciformes
Family: Mullidae
Length: maximum about 16 inches (40 cm)
Diet: small benthic animals
Longevity: about 10 years

Colour very variable, usually orange to red. Fairly heavy body; rounded head. Two short dorsal fins, forked caudal fin, ventral fins are level with the pectoral fins. Large barbels.

Two ways in which fishes have adapted to life on the sea floor are represented here by the sole (*above*), with its flat body and cryptic coloration, and the moray eel (*below*), a fearsome predator with a snake-like form.

are situated on top of the head and are thus directed straight upwards. Apart from this singularity these fishes benefit from a light and dark pattern of coloration, enabling them to blend perfectly with the surroundings as they lie motionless on the sea bed, half buried in the sand. Furthermore, the nostrils open in the mouth, an unusual situation among fishes, forming a vital part of an original respiratory system. The mouth is not used in the usual way for breathing, water entering through the nostrils and flowing out via the gills whereas in the general run of bony fishes water for breathing is gulped. The pectoral and pelvic fins have an additional use for walking and burrowing.

The astonishing stargazers have a veritable arsenal of weapons for use in self defence or in attack. In the first place they possess sharp venomous spines on the back; secondly, they have certain muscles behind the eyes which have been transformed into electric organs capable of giving out a shock of about 50 volts; thirdly, they are able to burrow rapidly into sand or mud up to a depth of some 12 inches; finally, when they go onto the offensive, they can attract prey by means of a worm-like filament with a fleshy projection at the tip. This is vibrated in front of the mouth and has the added advantage of being retractable, hidden under the tongue when not in use.

Another group of benthic bony fishes, similarly endowed to the stargazers, are the anglerfishes (Lophiidae) and the related frogfishes (Antennariidae) and batfishes (Ogcocephalidae), all of which are masters in the art of camouflage and employ a form of lure for fishing. The front part of the anglerfish's body is flattened and this tends to make it look rather like a carpet shark.

As with the cartilaginous group, there are numerous species of bony flatfishes. These belong to the huge order of Pleuronectiformes which include soles, turbots, plaice, flounder and their relatives. They are all well endowed to escape enemies, for the upper part of the body is capable of sudden colour changes according to the habitat concerned, thus blending harmoniously with the surroundings. These fishes lead a tranquil life, simply lying on the bottom; and since the body is flattened laterally they lie on their side rather than on the abdomen. When the alevins hatch from the eggs, their eyes are situated normally on either side of the head; but once they have grown to a certain size the fishes are strangely remodelled. One eye begins to move and eventually ends up on the same side of the body as the other eye. Simultaneously the fish wends its way down to the bottom, coming to rest on its blind side. Thereafter it spends its life in that position.

Other bony fishes that have adapted to benthic conditions are the Anguilliformes, including the conger and moray eels, powerful predators with snake-like bodies.

Dangers of inflation

There is only space to mention a few other fishes which have found highly original solutions to the problems of adjusting to life on the ocean floor.

The well known sea horses which, with the pipefishes, make

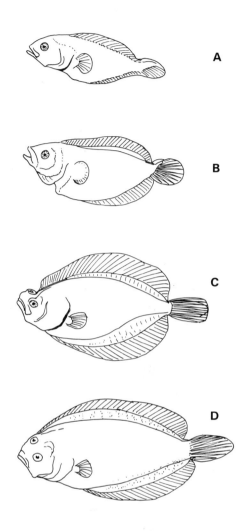

At birth the larvae of flatfishes of the order Pleuronectiformes are quite normal in structure. As they grow, however, one eye slowly shifts across the head to the opposite side. Eventually the underside becomes blind and this enables the fishes to lie flat on the ocean floor and retain the use of both eyes.

Many benthic fishes find their best means of self defence is to stay quite still. Pipefishes (*right*) blend astonishingly in shape and colour with the surroundings.

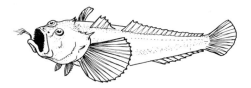

The stargazer (*above*) has a wide range of adaptations to benthic life—high-positioned eyes, pectoral and ventral fins that can be used as feet, a perfect pattern of camouflage, venomous spines, electric organs and a fishing filament or lure.

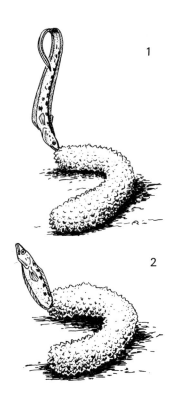

Pearlfishes often seek refuge inside sea cucumbers. The young enter head first through the anus, the adults tail first.

up the family Syngnathidae, swim in an upright position among fields of seaweed. Having scented food they literally suck it in with their long flute-like snout. A singular feature of the reproductive process is that during copulation the female sea horse deposits her eggs in an incubation pouch on the male's abdomen. The pouch is lined with a spongy tissue, similar to that of a placenta, for feeding the embryo. When the fry hatch the male expels them with a series of body contractions.

The pearlfishes of the family Carapidae live in the body cavities of sea cucumbers or, less frequently, inside the shells of bivalve molluscs. In the former case they live more or less as parasites, feeding on the sex glands; but the sea cucumbers are not adversely affected for they are able to eject all their internal organs, pearlfishes included, and also have strong regenerative potential.

The porcupinefishes (Diodontidae) and pufferfishes (Tetraodontidae) are covered with spiny projections and inflate themselves like balloons, by swallowing air or water, in self defence. Although most species are venomous, the effect of this elaborate defence mechanism may be delayed. A predator which swallows

one of these strange fishes is surely doomed, for the victim swells itself up inside the throat or stomach of its captor, taking on the guise of a huge stopper, thereby sealing vital passages. Even when dead the body remains inflated and the spines embedded in the predator's tissues. Some zoologists believe that the skin of pufferfishes and similar species is resistant to the action of digestive juices and that this upsets the predator's digestive system; true or not, the poison itself is likely to prove fatal. Other experts claim that the victims sometimes escape by biting their way out of their captor's body with their powerful, beak-like teeth.

Closely related to these species are the boxfishes of the family Ostraciontidae, but they are enveloped in a veritable bony suit of armour from which protrude only the eyes, the mouth and the fins. They appear to possess a similar defensive apparatus and their bright colours are clearly a warning that their flesh is poisonous. The fishes of all three families are very mediocre swimmers, propelling themselves with their dorsal and anal fins, feebly supported by the pectoral fins. The muscles designed for moving the tail are almost atrophied.

The strange boxfish, whose flesh is poisonous, is enveloped in a hard, bony carapace, with openings only for eyes, mouth and fins.

CLASS: Pisces

The first vertebrates to appear on earth—animals which possessed an internal backbone with distinct segments or vertebrae, supported by numerous organs and muscles—were the fishes. Modern zoologists include in the superclass of Fishes (formerly known as Pisces) all poikilothermal (cold-blooded) vertebrates which lead an aquatic life and which obtain the oxygen necessary for their vital functions by means of gills. They possess fins and the entire body is covered either with scales, denticles or bony plates. Such fishes are sometimes grouped together as Gnathostomata, literally referred to as 'jaw-mouthed'.

Although there is still considerable disagreement about classification, the representatives of the Cyclostomata or Agnatha ('jawless') are generally classified as fishes because, in spite of the fact that they do not possess a lower jaw, they have the characteristic fish shape. These include lampreys and hagfishes.

The first fishes appeared in the Ordovician period nearly 500 million years ago. These were primitive forms which reached their zenith during the Devonian period some 100 million years later. As time passed these ancient groups evolved and diversified. The 20,000 or so species of modern fishes are divided into four major classes.

The Acanthodii are fossil fishes dating from the Silurian and Permian periods, which were probably most abundant during the Devonian. They are the most primitive members of the superclass, being characterised by a partially ossified head, a non-bony vertebral column, a heterocercal (unequally divided) tail, and rudimentary fins that have been reduced almost to the form of spines.

The Placodermi are the jawed and armoured fishes, all of which are fossil species. They reached their peak in the Devonian period. The head and front part of their body were covered by large bony plates—this giving rise to their common name—but the rest of the body was either covered in small scales or was entirely naked.

The Chondrichthyes are the cartilaginous fishes whose principal feature is a cartilaginous internal skeleton, a skin covered with prickly denticles, and gill openings without an operculum (gill-cover). They have a wide modern distribution and are divided into two subclasses—the Elasmobranchii (sharks and rays) and Holocephali (chimaeras).

Finally there are the Osteichthyes or bony fishes which are the most highly evolved representatives of the superclass. In their case the skeleton is completely ossified, the body is covered with scales, and there is often a swim bladder associated with the digestive tube which fills with air and serves to regulate the hydrostatic equilibrium of the body.

The Osteichthyes are usually divided into four distinct subclasses. The Actinopterygii or ray-finned fishes comprise an enormous range of species, of varying shapes and sizes and with widely differing habits. The most primitive of these are the sturgeons and the most evolved the hakes, soles, etc. The Brachiopterygii, on the other hand, which include the bichirs, are descended from a group of very ancient fishes and are virtually an isolated branch. Of particular interest because of their rarity (little being known of their habits) are the Crossopterygii, represented by the recently discovered coelacanth. Study of their paired fins has enabled zoologists to understand the evolutionary processes by which fins were eventually transformed into the limbs of terrestrial mammals. Finally, the subclass Dipnoi includes unusual species, such as the lungfishes, capable of breathing air through lungs and therefore able to survive out of water.

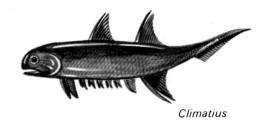
Climatius

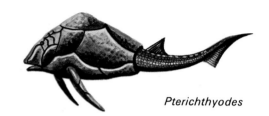
Pterichthyodes

Lesser spotted dogfish

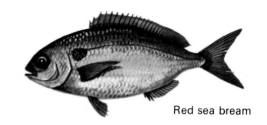
Red sea bream

The superclass of Fishes comprises four classes, as shown in the above drawings. The Acanthodii and Placodermi are represented by *Climatius* and *Pterichthyodes* respectively, both extinct, the Chondrichthyes by the lesser spotted dogfish, the Osteichthyes by the red sea bream.

Facing page: The majority of living fishes are Osteichthyes (bony fishes). The diagrams show principal external and internal features which have enabled such species to adjust to aquatic life. One such bony fish is the cod (*below*), a member of the Gadiformes.

CHAPTER 10

Cormorants, frigate birds and gannets

Although admittedly not scientifically correct to separate sea birds into three groups—linked respectively with the coast, the continental shelf and the open sea—there is much justification for considering them under these distinct headings since the inhabitants of each of these biomes possess a number of common structural and behavioural features.

Those birds which depend exclusively on the sea for food and which only visit dry land for nesting and breeding, form a biological entity quite distinct from species which find much of their food on land; and the former may, in turn, be conveniently divided into two groups—those that are pelagic, typical of the high seas, and those that frequent the shallower waters of the continental shelf. It is this second group that we shall be studying in this chapter.

Apart from the Alcidae of the order Charadriiformes, which, although essentially coastal birds are also found farther out to sea, all the species with which we are here concerned belong to the order Pelecaniformes. Of these the boobies, gannets, tropic birds and frigate birds are characteristic marine species, whereas the cormorants and pelicans are also frequent inhabitants of lakes and rivers.

Before describing the more important birds of the continental shelf it is worth referring to the authoritative work of Professor Jean Dorst, Director of the Paris Museum of Natural History, who has summarised the basic biological differences between the birds of the open sea and the birds of the shallower waters near the coast.

All the marine birds living fairly close to the shore, including both those of the coastal fringe and those of the continental shelf, generally have more food at their disposal than do the

Facing page: One of the most spectacular of sea birds is the magnificent frigate bird. In the breeding season the area of naked skin at the base of the male's neck turns red and swells to an enormous size.

Colonies of cormorants build bulky nests of branches and scraps of vegetation in rocky coastal regions.

species that are exclusively pelagic. Because animal life in these zones, which are bathed by the tides and illuminated by the sun, is so much more abundant and diversified than that which exists in the high seas, there is an equivalent wealth of bird species, all of which have a high birth rate. Food, both for adults and young, is freely available near the nesting sites, whereas species which fish far out to sea have long distances to travel to and from their nests. Everything is therefore in favour of the former groups, with the result that broods are larger and the growth of chicks much faster. Jean Dorst has pointed out that shags (birds of the continental shelf) leave their nests 55 days after birth, while Manx shearwaters (pelagic birds of roughly the same size) do this only after 70 days. Furthermore, and for the same reasons, the former are capable of reproducing at about three years of age (like gulls and other cormorants) whereas the latter are not sexually mature until they are five or six years old.

Information of this kind explains why population counts of coastal species fluctuate much more dramatically and rapidly than those of pelagic species, where the early mortality rate is considerably lower. In South America—to give one striking example—the numbers of guano-producing bird species are decimated from time to time, with literally millions of birds dying over a short period. In contrast, species such as albatrosses, which nest and give birth to a single chick only once in three years, remain relatively stable.

Apart from these differences affecting habitat, there are many physical, physiological and ecological adaptations which are to some extent common to all these birds, arising from their very conquest of the marine environment.

Oceanic birds, in fact, have had to cope with the same type of problem as faces all land vertebrates—the need to eliminate the salt that they take in while feeding in the sea, given the fact that the kidneys of animals which have evolved on land are incapable of functioning if the salt content of imbibed liquid exceeds a certain limit. Should a man, a dog or any land bird try to quench its thirst with sea water it will be even thirstier afterwards, for the kidneys are forced to expel a large quantity of water in order to dissolve the sodium chloride which it contains.

How then do marine birds, which do not drink any fresh water, get rid of this unwanted salt? It was once thought that they overcame this problem because of specially modified kidneys; but investigations have now shown that the kidney tissue of such species differs hardly at all from that of land birds. Studies by Schmidt-Nielsen have demonstrated that sea birds eliminate salt by means of certain glands (so-called 'salt glands') situated in the upper part of the eye socket. These glands, semi-circular in shape, are less or more developed according to the measure of adaptability of the particular species to ocean life. Obviously they are most efficient in the pelagic species. The glands, both in structure and function, bear a resemblance to true kidneys, complete with small veins, arteries and tubules.

Within minutes of pure salt being fed to a Humboldt penguin, a clear colourless liquid started to drip from the tip of its beak.

The experiments, designed to test how extraordinarily efficient the salt glands of seabirds are, were conducted by the American biologist Schmidt-Nielsen, on a penguin weighing about 13 lb. Only five grammes of salt was used, dissolved in water. Incidentally, similar salt glands are found in gulls and other seabirds, as well as in seals and turtles. After about four hours, examination of the liquid emitted in this way showed it to contain approximately two-thirds of the salt that had originally been administered. The result of the experiment was significant in demonstrating how effectively the salt glands perform their work of filtration, bearing in mind the fact that, over the same period, a kidney is capable of eliminating only about one-tenth of the salt contained in sea water.

Another vitally important attribute which has helped such birds to come to terms with their ocean environment is, of course, their amazing capacity for flying, this being particularly evident among pelagic species. The land birds which are acknowledged to be the most remarkable fliers, such as the members of the stork family and the various birds of prey, take every possible advantage of rising currents of warm air. These thermal currents carry them to considerable heights, from which vantage points they can make long angled dives on their objectives. Warm air currents, however, are not in evidence out at sea; consequently migrating species try to avoid excessively long journeys over water, crossing from one continent to another by straits (such as Gibraltar) or an isthmus, such as that of Sinai. There is evidence that where birds, and especially the small songbirds, are forced to make an extended sea crossing they are apt to incur many casualties.

To cross an ocean a bird must either be able to fly in the normal manner or possess a specialised wing structure that enables it to glide over the water. Standard flight, involving series of steady wing-beats (as is adopted, for example, by plovers in the course of their fantastic migrations) is quite clearly impractical—because it wastes too much energy—for normal daily activities such as fishing. It is therefore for purely economic reasons that many sea birds, particularly albatrosses, shearwaters and petrels, are endowed with specially structured wings, including long, narrow, pointed remiges which make it possible to climb and make steady forward progress against the wind without being carried off course or losing too much altitude. Storms and hurricanes are no deterrents. All the birds have to do is to keep altering the angle of incidence of their wings, adjusting to the speed and direction of the wind, expending the least possible energy. This explains how those prodigies of gliding flight—the albatrosses—manage to circle the globe, only visiting dry land once every three years for breeding.

Living more or less permanently at sea obviously entails an aptitude for obtaining food from the water and normally this involves swimming. But the species most adept in swimming and diving have either lost their flying powers completely or are relatively poor fliers over any distance. In fact the wing serves either as a flight apparatus or as a swimming mechanism, rarely both. Under water many birds keep the wings folded, using

British ornithologists have made a study of the courtship ceremonial of cormorants and gannets. Cormorants and shags go through a special sequence of movements as they prepare to incubate the eggs (1) and (2). Gannets and boobies throw up their head to reveal the black patch of naked skin on the throat in order to restrain undue aggressiveness in another bird of the same species (3) and (4).

their powerful webbed feet for propulsion. Cormorants provide striking examples of this procedure. Others, such as pelicans, use the sheer inertial force of their falling bodies in diving from various heights. Some species, including petrels, fly so low over the water that they almost appear to walk on the surface. But the simplest method is that employed by albatrosses and related species which bob about on the waves and capture prey with their long hooked beaks.

Among the most accomplished diving birds, with wings transformed into flippers, are penguins which, like the extinct great auks, cannot fly at all; other Alcidae (puffins, little auks, guillemots and the like) 'fly' under water and also use their wings for short flights through the air. Of the pelagic birds only the diving petrels (Procellariiformes) use their wings both for swimming and flying; but they are incapable of gliding in the manner of other petrels and albatrosses.

Geographical distribution of common cormorant.

Cormorants and shags

The rocky coastline of Cantabria in northern Spain, battered at the height of winter by furious waves, provides the backdrop to one of the most remarkable spectacles in the bird kingdom. Perched immobile on the islets dotted along the shore are hosts of blackish birds, their plumage glistening with a metallic sheen. Imperturbably they stand, like ranks of sentinels, defying the elements. Colour and shape identify them unmistakably as cormorants.

Not every member of this impressive colony is content to stand and watch the turbulent ocean. Out at sea there are black shapes floating low on the crests of the waves, following the rising and falling swell, then leaping forward and vanishing from view. Although invisible, they are now in their true element. Propelling themselves downward with their strong webbed feet, they pursue their prey, remaining below for perhaps a minute, then veering up again to the surface.

The various species of cormorants and shags are among the most expert fishing birds anywhere, hunting (according to individual choice) with equal ease and proficiency in rough seas, tranquil bays, estuaries, lakes and rivers. Their fishing skill is so extraordinary that in the Far East men have exploited them ever since ancient times for this very purpose, trained cormorants being attached to lines held by the fishermen and prevented from swallowing their catch by a leather collar slipped around the neck. Fishing with cormorants is still practised in Japan, partly for the benefit of tourists; and on the French island of Orleron it has been organised as a sporting attraction, the participants using modern diving equipment to enjoy the spectacle of the graceful birds under the water.

The family Phalacrocoracidae contains about thirty species of cormorants, some of which are known in certain parts of the world (sometimes interchangeably) as shags. They are generally found in coastal regions but some species inhabit inland lakes and rivers as well. The family is subdivided into two, or according to certain authors, three genera. Most species belong to the

---

CORMORANTS

Class: Aves
Order: Pelecaniformes
Family: Phalacrocoracidae

COMMON CORMORANT
(*Phalacrocorax carbo*)

Total length: about 38 inches (95 cm)
Weight: male 5 lb (2.3 kg)
female 4¼ lb (1.9 kg)
Diet: fishes
Number of eggs: 3–4
Incubation: 28–31 days

Blackish plumage; chin and cheeks white. In breeding season there are white marks on the thighs and around the throat. Back and belly of young are also white.

SHAG
(*Phalacrocorax aristotelis*)

Total length: about 30 inches (75 cm)
Weight: 4¼ lb (1.8 kg)
Diet: fishes
Number of eggs: 3
Incubation: 24–29 days

Black plumage with green or blue reflections. Head crest in breeding season. Smaller than common cormorant, with more slender beak. No white marking on head.

PYGMY CORMORANT
(*Phalacrocorax pygmaeus*)

Total length: 19 inches (48 cm)
Weight: about 1½ lb (700 g)

Very small bird. Breeding plumage brownish-black, streaked with white, slightly green on head and neck; beak shorter and thinner than that of other species.

genus *Phalacrocorax*, including the widely distributed species (very familiar in Europe) known as the common or great cormorant.

The plumage of cormorants is dark brown or black, with green, blue, purple or bronze tints. The wings are comparatively short and the tail, composed of stiff feathers, serves as a rudder under water. The bill, with sharp cutting edges, is slender and hooked. The powerful feet, placed well to the rear of the body, terminate in four webbed toes. Although graceful in water, the birds are extremely clumsy on land, walking with a comic waddling gait; but the long prehensile toes help them to keep their balance on any kind of hard surface, including the slender branches of tropical trees, slippery spurs of rock or even the masts of sailing ships.

An unusual feature in a bird which is so magnificently suited to life in water is that the cormorant possesses no oily uropygeal secretions which are of such value to other aquatic birds for preening and waterproofing the feathers. Once wet, therefore, the body of the cormorant becomes much heavier. This has its advantages and disadvantages. In the water it helps the bird to dive deeper and more rapidly for fishes; but on dry land it is compelled to remain quite still, wings outspread, so that the water can evaporate. The outspread wings also help to keep the birds spaced when on land. Flight, at any time, is a ponderous affair for most species, particularly when emerging from water, this entailing extra-strong thrusts with the legs. Once airborne the bird flies with neck outstretched, beating its wings vigorously, somewhat in the manner of a wild duck, but with a much less elegant effect.

The absence of land predators (as a result of geographical isolation), plus an extremely specialised aquatic existence, have caused the flightless cormorant of the Galapagos Islands to lose its flying powers entirely. The wings of the species are rudimentary and there is no keel to the breastbone. The two guano-producing cormorants, however, are good fliers.

Although cormorants supplement their diet occasionally with crustaceans and (in rivers and lakes) amphibians, they feed principally on fishes. These are caught underwater with the aid of the strongly hooked beak. If large they are brought to the surface; if small they are swallowed immediately.

There are two species of cormorant in western Europe. The common cormorant (*Phalacrocorax carbo*) is roughly the same size and weight as a gosling and nests on coastal cliffs, in trees near the sea or on the banks of lakes and marshes. The shag or green cormorant *(Phalacrocorax aristotelis)* is smaller and only to be found on rocky coasts. Apart from their size the two species can be distinguished in other ways. During the breeding season the dark plumage of the common cormorant is adorned by two white patches, one at thigh level, the other fringing the throat. At the same season the shag (known in parts of Europe as the crested cormorant) displays a small tuft of feathers on its head. Feeding habits also differ. The former catches both sea and freshwater fishes whereas the latter only hunts species commonly found on or around rocky sea beds.

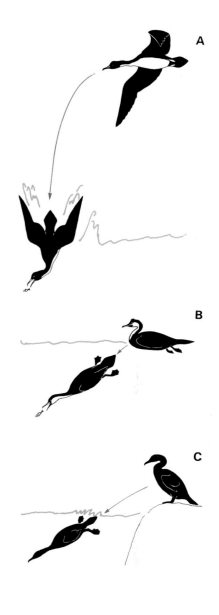

Cormorants, as a rule, are not good fliers, but are remarkable for their diving and underwater fishing techniques. But methods of hunting vary according to species. Thus the guanay cormorant (A), with long, pointed wings, hovers over the sea and swoops on its prey close to the surface. The shag (B), with only average flying ability, like the majority of species, dives from the surface, propelling itself to greater depths with its webbed feet. The flightless cormorant (C) has lost its flying powers completely.

Facing page: Most cormorants nest and breed on cliffs overlooking the seashore or on rocky islands. The species shown here are pelagic shags of the Pacific (*above*) and the common cormorant (*below*).

Geographical distribution of magnificent frigate bird (*shown below*).

MAGNIFICENT FRIGATE BIRD
(*Fregata magnificens*)

Class: Aves
Order: Pelecaniformes
Family: Fregatidae
Total length: 40½–44 inches (103–112 cm)
Wingspan: 79–91 inches (200–230 cm)
Weight: 3–4½ lb (1.4–2 kg)
Diet: fishes and sea animals
Number of eggs: one
Incubation: 40–50 days.

Black plumage flecked with white. In breeding season male has bright red, inflatable throat pouch. Long bill, curved at tip. Very large, narrow, pointed wings. Other species similar.

Common cormorants often form large colonies. They reach sexual maturity between three and five years of age and the females lay three or four bluish-white eggs. Incubation lasts approximately one month and the chicks leave the nest after about seven weeks, although remaining dependent upon their parents until they are three months old. Female shags lay only three eggs and in their case the incubation period is 24–29 days. The chicks are rather more precocious but they too spend some months with the parents.

The pygmy cormorant (*Phalacrocorax pygmaeus*) is much smaller than either of the afore-mentioned species and lives only in eastern Europe. It breeds in colonies among reeds, on low shrubs or in clumps of ash trees. Females lay from three to six eggs.

Frigate birds and tropic birds

Ever since men have navigated tropical seas they have been much attracted by marine birds. One reason is that they form a pleasant distraction from shipboard duties, some of them being strikingly graceful and beautiful in flight. Another probable reason is that their presence is often the first sign that land is near. Of such species the frigate birds and tropic birds are both, in their own ways, remarkable.

The frigate birds of the family Fregatidae, sometimes known to sailors as man-o'-war birds, are splendid fliers. They seem to be carried effortlessly by the wind, soaring high above the sea for hours on end, occasionally changing direction or gaining speed with a series of long, slow wing beats. Hanging almost motionless in the blue sky they rarely settle on the water. The tropic birds (also called bo'sun birds) of the family Phaethontidae, on the other hand, with their compact streamlined body extended by a pair of central feathers up to two feet long, fly with rapid wing beats, rather like falcons. Sometimes they plummet down out of the blue, like boobies, snatching up fishes with their conical pointed bill when only a few inches above the surface.

Tropic birds and frigate birds differ greatly in appearance and behaviour. Whereas the white tropic birds, flapping their wings, catch fishes directly from the sea, the black frigate birds are champions of gliding and obtain most of their food by chasing other sea birds, tormenting them mercilessly until they drop their prey.

The five species of frigate birds, all belonging to the genus *Fregata*, possess long pointed wings, a sharply forked tail and short legs. Although extremely awkward and clumsy on land, once in the air they are a marvel to watch. Ornithologists consider them the most aerial of all sea birds. They spend most of their life hovering high over the ocean although seldom straying far from land.

Frigate birds are true aerial pirates, living at the expense of other species. Gulls, cormorants and, above all, boobies are chased and pecked repeatedly until they disgorge or drop their food. This is recovered in mid-air before it reaches the water. Sometimes frigate birds swoop down on breeding colonies of

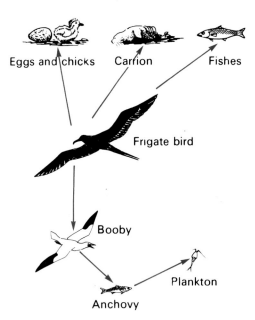

In addition to eggs and chicks of other sea birds, frigate birds feed on fishes, invertebrates, baby turtles and carrion. They also harass boobies, pecking at them until the latter disgorge their food.

Frigate birds and tropic birds have recognisable shapes and colours. The former (*above*) are predominantly black with a prominently forked tail; the latter (*below*) are mainly white, with two long streaming rectrices.

marine birds and make away with eggs and chicks. The long hooked beak is also ideal for plucking baby turtles or fishes out of the sea as the birds skim along the surface, hardly wetting their feathers.

All five species of frigate birds have blackish plumage but the males are distinguished during the breeding season by a bright red throat pouch (normally orange) which is not present in the females. The rudimentary nest of branches may be situated in a tree, in a bush or on the ground. The male initiates his courtship display by spreading his wings and inflating his crimson throat pouch until it reaches truly astonishing dimensions, completely dwarfing his head. By reason of its size and brilliant colour such a visual signal cannot possibly be ignored either by the female whom it is designed to attract or by a rival who may be planning to usurp the nest.

Frigate birds nest in colonies alongside the boobies which so frequently provide them, albeit unwillingly, with food. In those areas where fishing conditions are good there seems to be fierce competition among pairs of frigate birds for the choicest nesting sites. The female lays only one egg which is incubated by both parents in turn, and the chick hatches within 40–50 days. It is not completely covered with feathers, however, for four or five months.

The various species are well dispersed, three of them with a wide range. The magnificent frigate bird (*Fregata magnificens*) is a New World species, found in the Caribbean and along the coasts of Central and South America; the great frigate bird

The red-billed tropic bird finds shelter in fissures of rock. In order to avoid damaging its splendid tail it settles face-inwards, leaving the long tail feathers hanging out.

Facing page: The frigate bird often pillages the nests of other sea birds (*above*). Its own chick is born naked but acquires a covering of white down within a couple of weeks (*below*).

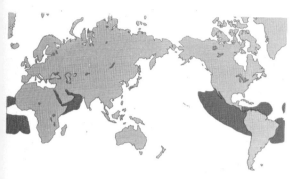

Geographical distribution of red-billed tropic bird.

RED-BILLED TROPIC BIRD
(*Phaethon aethereus*)

Class: Aves
Order: Pelecaniformes
Family: Phaethontidae
Total length: about 32–40 inches (80–100 cm) including tail feathers
Diet: fishes and sea animals
Number of eggs: one
Incubation: 41–45 days

Plumage white, spotted and striped with black. Head narrow, bill pointed. Wings long and slender. Two extremely long tail feathers.

(*Fregata minor*) roams the Indian Ocean, the Central Pacific and the south-western Atlantic; and the lesser frigate bird (*Fregata ariel*) breeds north of Madagascar, off the coast of Brazil and on islands in the South Pacific. The Ascension frigate bird (*Fregata aquila*) and Christmas Island frigate bird (*Fregata andrewsi*) have a more limited range, in the South Atlantic and Indian Ocean respectively.

There are three species of tropic birds—the red-billed tropic bird (*Phaethon aethereus*), the red-tailed tropic bird (*Phaethon rubricauda*) and the white-tailed tropic bird (*Phaethon lepturus*). All have the appearance of being more closely related to gulls and terns but are in fact allied to boobies, cormorants and pelicans. Plumage is basically white, with a sprinkling of black marks on head and wings. The red bill and tail of the two species so named are identifying features, and these are also larger than the white-tailed tropic bird. Food consists of fishes and squids which are caught near the surface as the birds plummet down from a height of 50 feet or so, emerging after a few seconds with their prey.

Tropic birds nest in colonies among rocks. The chicks are born covered with down (like baby gulls) and are fed by their parents until they have acquired a conspicuous layer of fat, when they are ready to leave the breeding site and start fending for themselves.

At one time tropic birds were hunted for the sake of their handsome tail feathers; but today the inaccessibility of the breeding sites in tropical seas is a reasonable safeguard.

The spectacular gannets

The northern gannet (*Sula bassana*) is a beautiful species and undoubtedly one of the most spectacular diving birds of European and North Atlantic waters. The sight of such a bird dropping into the sea like a stone is a reliable signal to other species that there is an appetising shoal of herrings in the vicinity, a clue equally welcomed by skippers of trawlers.

Acquiring such an accomplished fishing technique has entailed a number of physical modifications. Since the gannet dives with folded wings from 100–150 feet there is, first and foremost, the problem of cushioning the body against the great shock of impact with the water. The presence of a network of air sacs below the skin helps considerably; and because the nostrils are sealed, that is they are occluded and breathing is through the mouth, water cannot get in to interfere with breathing.

The nine species of gannets and boobies all belong to the family Sulidae and the single genus *Sula*. The six boobies (so named years ago because of their alleged stupidity in allowing humans to catch them and raid their nests) dive and hunt fishes underwater in much the same way as gannets, but are birds of tropical seas. The other gannet species are the Cape gannet (*Sula capensis*) of South Africa and the Australian gannet (*Sula serrator*). All are predominantly white with black tail and wing tips, but two of the boobies are brown. The chicks are dark brown and take several years to acquire their adult livery. The birds are stoutly built with a large head, a thick neck and a long

straight bill. The four toes are webbed. Since the nostrils on the bill are non-functional the problem of breathing is resolved by air entering through the notched edges at the base of each half of the beak, this being held half-open to allow sufficiently free circulation.

Although expert at flying and swimming, boobies and gannets are very clumsy on land and only leave the water in the breeding season when they gather in their thousands, building bulky nests of seaweed and grass. Each female lays one egg and the naked chick is fed with regurgitated food.

Guano: a natural treasure

On the west coasts of South America and South Africa a combination of geographical, climatic and ecological factors have contributed to a remarkable natural phenomenon which may well have begun millions of years ago but which has been discovered and exploited by man only in fairly recent times. The droppings of innumerable sea birds such as pelicans, cormorants, boobies and penguins, preserved by a complex sequence of interacting factors, have been converted into a natural treasure of inestimable value – guano. The word is of Spanish derivation, borrowed from the Inca tongue to describe the substance, rich in nitrogen, excreted by these sea birds. The term has since come to be applied to the droppings of other vertebrates, generally cave-dwelling species such as bats; and this product has proved to be of great commercial value as a fertiliser.

Two members of the family Sulidae are the red-footed booby (*left*) and the northern gannet (*right*). Both are excellent fliers and divers, though awkward on land.

As in the case of other Pelecaniformes, the nostrils of boobies and gannets are sealed. Breathing is effected through notches along the edges of each half of the bill. Air circulates freely when the beak is kept half-open, and there is no risk of water entering the nostrils when the bird dives.

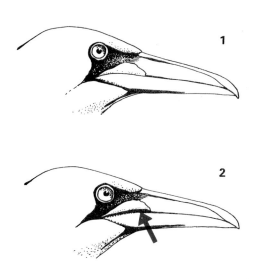

Geographical distribution of northern gannet.

NORTHERN GANNET
(*Sula bassana*)

Class: Aves
Order: Pelecaniformes
Family: Sulidae
Length: 36 inches (91 cm)
Wingspan: 68 inches (173 cm)
Weight: up to 6½ lb (3 kg)
Diet: fishes
Number of eggs: one
Incubation: 42 days

White plumage; black primary rectrices. Long, straight wings; pointed bill; narrow tail. Young have brown plumage with white marks, only becoming paler after some years.

Facing page: During the breeding season Cape gannets assemble in immense colonies on certain small rocky islands off the coasts of South Africa. Their droppings, containing a large amount of nitrogen, form thick layers of a substance known as guano, valuable as a fertiliser.

The highest quality guano is found in South America and South Africa. Although available as well on many oceanic islands, the high level of humidity in these areas brings about important chemical changes in the composition of the substance so that it loses a large proportion of nitrogen whilst gaining in phosphate content. For commercial purposes this guano, though usable, is of less value than the type collected in the two principal production zones.

The presence of such enormous colonies of sea birds and hence of vast deposits of guano in these regions is the result of a complex interplay of forces, two of which may be singled out as being particularly influential. One relates to the properties of the ocean itself, the other to climate. The physical structure of the seas is partially responsible for the strong currents which flow in the two areas concerned – the Humboldt, off the shores of Chile and Peru, and the Benguela, along the coasts of West Africa. These currents, continually stirring up water in the depths, drive minerals to the surface. As a result these zones are especially rich in phytoplankton and zooplankton, providing abundant food for huge shoals of anchovies. Millions of sea birds are attracted by the teeming shoals and form tightly packed colonies on the islands off the coasts, safe from land predators. The droppings heap up, growing some three inches year by year, and because of the extraordinarily dry climate the piles of excrement, with their high nitrogen content, may be preserved, unchanged in composition, for centuries.

Scientists who have examined such guano deposits, taking into consideration the present height of the heaps and the estimated rate of annual growth, have concluded that in Peru their formation dates back to at least 500 B.C. But the use of more accurate techniques on certain sites indicates that they must be considerably older than that, probably dating from the Pleistocene epoch, more than two million years ago. The continuity may, however, have been interrupted because of climatic fluctuations, as the islands were subjected alternately to periods of drought and heavy rain.

There is conclusive evidence that guano was already used in Peru in prehistoric times and that the practise still continued in 1609, the publication date of Garcilaso de la Vega's *Royal Commentaries of the Incas*, in which he described how the local Indians regularly collected the material. After the conquest of Peru the guano industry continued to flourish but it was another two centuries before the German explorer Alexander von Humboldt revealed its true commercial potential.

Judging from reliable reports it was during the years following 1840 that guano began to be exploited on a really large scale. Whereas the Incas had been careful to protect the birds and their breeding sites, the 19th-century colonists, eager for quick profits, ransacked the islands without scruple. According to R. C. Murphy, some 20 million tons of guano were exported from Peru between 1848 and 1875.

The thorough and relentless destruction of all these bird colonies inevitably led to near exhaustion of local guano supplies. Around the beginning of the present century the situation was

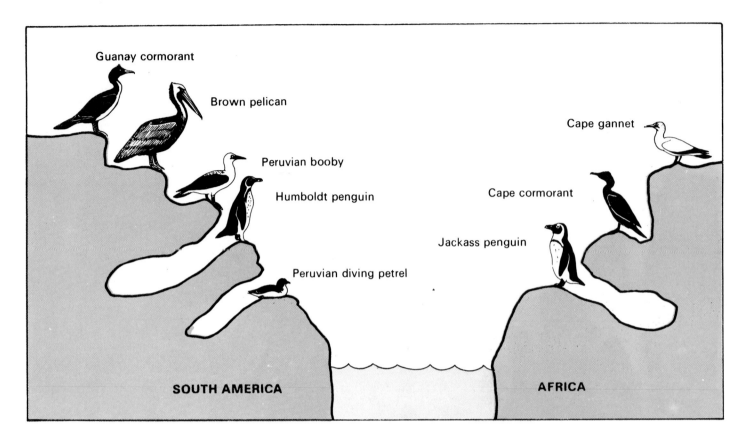

There are obvious affinities between the guano-producing sea birds of South America and South Africa (*above*). Thus the Peruvian or Humboldt penguin (*below*) and the jackass penguin belong to the same family and play the same ecological role. There are no African equivalents, however, of the brown pelican and Peruvian diving petrel.

so serious that Peruvian agriculture was gravely threatened for lack of the precious fertilising material. In 1909, therefore, a semi-official organisation was set up to look into the entire problem, salvaging what it could and laying the foundations for a healthier future on a rational cropping basis. The results were slow but promising. The birds gradually returned to the sites from which they had been driven and even settled on cliffs that they had previously ignored, rewarding the efforts of conservationists who had established 'reception areas' surrounded by barbed wire which proved to be extremely effective against poachers. The annual figure of 20,000 tons of guano which had been extracted at the end of the 19th century had risen to more than 300,000 tons a year by 1960.

Reconstitution of the bird colonies was not, however, accomplished without problems. Peru had in the meantime developed into the world's leading supplier of anchovies. Much of the catch was being used in the manufacture of fishmeal, itself a fertiliser, and flour. The difficulty was that anchovies were also the principal food of the guano-producing birds. Intensive study showed that only one-tenth of the food eaten by the birds was recovered in the form of guano whereas the output of the factories producing artificial fertilisers was twice as great. Fortunately the quality of the natural guano was much higher and its cost cheaper than that of the manufactured variety, so that the balance swayed once more in favour of the birds. Should it happen, nevertheless, that one day it proves possible to turn out a product competitive on all counts, there is a grave risk that the cormorants, gannets and pelicans will once more be endangered by being robbed of their food and that this time there may be no saving miracle.

Populations of guano-producing birds are subject to wide fluctuations due to purely natural causes. Thus even slight changes in climatic conditions may lead to catastrophe. Every seven years, for example, a warm counter-current from the north sets in motion a chain reaction which results in a sharp decline in numbers. The mineral content of the water drops and this affects plankton production. Less plankton means less anchovies and less food for the birds. Furthermore, colonies are often decimated by tuberculosis epidemics. As birds die in thousands, survivors migrate to other regions.

In South America the principal guano-producing species are the brown pelican (*Pelecanus occidentalis*), the guanay or Peruvian cormorant (*Phalacrocorax bougainvillei*) and the Peruvian booby (*Sula variegata*). All these birds build nests on top of the soft layers of guano and leave more excrement around the breeding sites. Other birds, also scooping hollows in the guano for nesting purposes, contribute their share to the total quantities. They include the Humboldt or Peruvian penguin (*Spheniscus humboldtii*) and the Peruvian diving petrel (*Pelecanoides garnotii*). These two species were especially affected by the mad rush for guano at any cost which occurred in the 19th century. In the course of literally stripping the rocks bare, countless nesting sites were demolished; and even when the adult birds managed to survive they could not be induced to return to breed the following year because there was no guano left on which to build their nests.

In the breeding season the three main guano-producing species form what are undoubtedly the largest and most crowded colonies in the world. When their eggs hatch and the young are born, the numbers of sea birds huddled together on the various islands off the coast of South America run into millions. It is estimated that there are at least three nests to every square yard of ground.

It is interesting to note that the brown pelican, Peruvian booby and guanay cormorant, are all surface fishers, specialised in catching anchovies. Normal though such tactics may be for the first two species, it is unusual for a cormorant. But this bird, in contrast to most other cormorants (which dive from the surface) locates its prey whilst in flight and then swoops down for the kill. More aerial by habit than related species, the guanay has long and powerful wings. The back is black, the breast white and a small crest adorns the top of its head. The bird's feet are red and there is a ring of naked skin around the eyes.

The guano-producing birds of South Africa have to contend with the same problems as those from South America. Fortunately, after years of ruthless exploitation, a more sensible, controlled cropping policy has been introduced, with effective protection for the birds involved. There are three species–the Cape cormorant (*Phalacrocorax capensis*), the Cape gannet (*Sula capensis*) and the jackass penguin (*Spheniscus demersus*). All of them have a restricted habitat on the south-west coast of the continent. Like its New World counterpart, the Cape cormorant is a strong flier and fishes in the same way.

Apart from anything else, this provides yet another example of convergent adaptation.

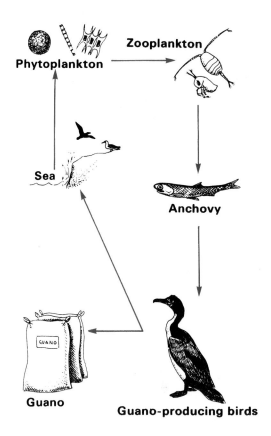

The guano cycle is a good example of the way in which the interaction of different factors determine the structure of a food chain. Ocean currents, by stirring up phosphates, stimulate the growth of phytoplankton and this makes possible the development of enormous quantities of zooplankton, food of anchovies which are in turn eaten by guano-producing birds. Guano is converted into agricultural fertiliser but some of it is washed back by rain into the sea, adding to its mineral content.

CHAPTER 11

The intelligent, friendly dolphins

Throughout the ages there have been charming references to dolphins, with particular emphasis on the sociable nature of these aquatic mammals. The writers of ancient Greece often described the manner in which dolphins would actually befriend sailors in distress and this theme was echoed in later legends of the sea. Such stories have usually been dismissed as delightful fantasies; but a number of incidents have recently occurred which give cause to wonder whether the ancient tales may not have contained a vestige of truth after all.

There was, for example, the famous case of Pelorus Jack, a dolphin which used to 'pilot' ships plying regularly between certain New Zealand ports for 20 years from 1888. Another less familiar episode was reported from Spain where a female dolphin (erroneously given the name of Nino) struck up a friendship with a professional diver named Antonio Salleres. Until her accidental death this animal would frolic happily with swimmers in a bay off the Galician resort of La Coruña, but continued to show special affection for her first friend.

There are several reports of dolphins coming to the assistance of bathers in trouble. In 1970, for example, Yvonne Bliss was rescued by a dolphin on the east coast of Grand Bahama, which nosed her ashore. This astonishing behaviour may be more a matter of instinct than intelligence, for observers have on a number of occasions seen dolphins of various species coming to the aid of injured companions. Yet this explanation is not fully satisfactory because there is a great difference between a dolphin helping another animal of its own kind and actually guiding a human to safety on dry land. The latter would seem to imply intelligent comprehension of a situation far outside the dolphin's normal range of experience.

Facing page: Dolphins are proverbially friendly, sociable animals. In the ocean dolphins of different species often group together to feed; and the same tendency appears among dolphins in aquaria, as shown here. The animal on the left is a bottle-nosed dolphin, the other a Pacific white-sided dolphin (*Lagenorhynchus obliquidens*).

Dolphins live in social groups but because they are so difficult to study in the wild little is known about the structure and numbers of such communities.

Since the Second World War scientists have spent much time studying dolphins in captivity and have been struck by their gentle attitude towards keepers and trainers. After being caught in nets they are cushioned in foam rubber so that there is no risk of suffocation when exposed to the force of gravity. Even at this point the animals remain calm and do not attempt to struggle, as if aware no harm is intended. It is interesting to note too that dolphins appear to be even better disposed towards children than adults. The reason may be that they respond to the same types of parental stimuli or that the playful tendencies of children correspond more closely to their own natural behaviour patterns.

An evolutionary marvel

The dolphin is one of nature's masterpieces. From a land mammal, walking and breathing air, evolution has fashioned an animal that has adapted to life in the sea so successfully that it rivals and even excels the fishes. To achieve this the entire organism of the cetacean has been transformed. In the words of Professor Grassé, '... there is not a single organ in its body that has not been modified as a result of its particular mode of life ... Issue of an evolutionary process that modern scientific theory is incapable

of explaining, the dolphin offers for the consideration of the research biologist the most extraordinary range of adaptations that can be imagined...'

One of the most surprising yet least noted of all these adaptations concerns its astonishing swimming ability. It has been calculated that a dolphin is capable of moving through the water at remarkably high speeds – up to 15–20 miles per hour – which seems inconsistent with its very average muscle power. If one were to try to construct a model of a dolphin, no matter of what kind of material, which was able to cut through water as rapidly as the animal itself, a vast amount of energy would have to be utilised. Indeed a dolphin is reckoned to have ten times the power of a man-made submarine. This is due to two important factors – the hydrodynamic outline of the animal and the unusual structure of its skin. The body is completely smooth but the amazing thing is that when swimming at high speed the dolphin is hardly affected at all by water resistance, which is the impeding factor to the progress of ships and other floating bodies. The skin plays a significant role, for it is pliable and takes up the correct shape to induce laminar rather than turbulent flow thereby reducing drag. It was only when dolphins were kept in aquaria that this was discovered.

It has been established that dolphins swim by means of broad

Most of the observations of dolphins have centred on the bottle-nosed species. Since they feed on benthic rather than pelagic fishes they tend to be rather sedentary and do better than others in captivity.

vertical movements of the tail (consisting of two fleshy lobes or flukes), their passage through the water being so clean that they leave no trail in their wake. Yet the real source of their amazing motive power remains a mystery.

Dolphins have also managed to solve the problem of maintaining a stable body temperature, higher than that of the water. A thick layer of subcutaneous fat or blubber efficiently insulates the body and prevents too much heat being lost; and the fact that the skin is poorly provided with veins and arteries is an additional factor reducing heat loss. But insulation poses problems, for when a dolphin exerts violent physical effort, the muscular activity involved produces a certain amount of heat which has to be released. This is effected with the aid of the fins, for these have a thinner covering of fat and a more adjustable circulatory mechanism than the rest of the body. Thus if the animal, when travelling at high speed, needs to lose heat, circulation in the fins is accelerated so that the blood quickly cools. But if the dolphin is resting or moving slowly, the blood flow is stemmed so that the temperature remains constant, the body losing only a small amount of heat to compensate for the little produced.

Channels of information

Dolphins, like other animals, have need of a continuous flow of information as to what is happening around them; and they obtain such information in very special ways.

None of the Odontoceti has a sense of smell, so far as we know. Certainly those investigated have always proved to have the olfactory cell atrophied, so they are truly anosmatic. The Mysticeti, by contrast, have an olfactory nerve. Their vision too is mediocre, although it varies according to species. Despite the fact that in their optic system the index of refraction is adapted to underwater vision, it must be remembered that in this world little light penetrates. Outlines and shapes soon become blurred or fade from view. Divers often experience a sense of disorientation, the feeling of being suspended in a void (aggravated by weightlessness), so that it is impossible to measure distances with any accuracy or even to distinguish the surface from the bottom. This so-called blue wall has been the cause of many an accident; and experienced divers know that the most reliable directional guides are air bubbles rising to the surface.

Because their eyes are adapted to the index of refraction of sea water, dolphins appear short-sighted when outside their natural element. As a rule they can see moving objects at a maximum distance of about 50 feet. Yet the fact that they can jump through hoops with great precision and respond to the slightest gestures of their trainers is an indication that many of them are capable of overcoming this handicap.

It may seem sad to think of dolphins as half-blind animals that maintain little visual contact with their surroundings; but in fact they do not need sympathy on that score, for they live instead in a world of sounds. Experiments have shown that their hearing mechanism is of great complexity and more highly perfected than in any other animal species.

The auditory mechanism of dolphins is far more complex than that of humans for it is an active rather than a passive process. Although this was long suspected, it was W. N. Kellogg who first proved conclusively that dolphins used an echolocation system, far more efficient than any sonar apparatus employed by surface vessels for the detection of submarines.

Dolphins (and cetaceans in general) emit sounds which in many cases are beyond the range of the human ear, by vibrating the external folds and internal cavities associated with the spiracle or blow-hole which is situated in the top of the head. When these series of rhythmic impulses strike an obstacle they are reflected back and reach the animal's inner ear. The receptor mechanism thus appears to be somewhat different from that of most animals inasmuch as the sound does not enter via the external ear (which among dolphins is a barely visible opening) but is evidently picked up by the lower jaw and transmitted by a network of nerve fibres ending in four holes in the bone.

Much more complicated than any man-made electronic apparatus used for locating submarines, the sonar mechanism of dolphins is still not thoroughly understood. It is quite clear,

Dolphins breathe through a single blowhole in the top of the head, closed by a special muscle when the animals dive. The mouth plays no part in respiration.

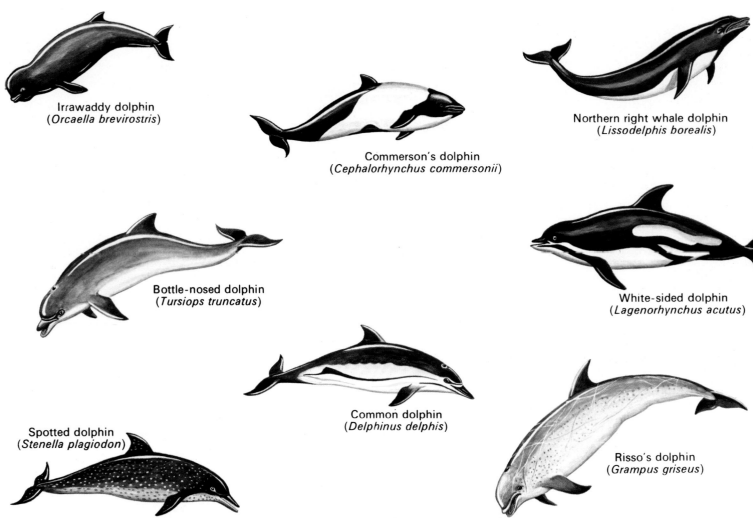

however, that it is capable of providing the animal with a wide range of extremely valuable information concerning the nature of underwater objects. Thus the variations in frequency of the emitted sounds enable the dolphin to obtain a number of echoes, comparable to a series of photographs of the same object taken in different colours or with varying filters. In an identical manner, different types of emission serve several functions. Thus certain sounds may convey general but unspecific information, simply giving warning, for example, of the presence of some kind of object. Once this object is pinpointed another kind of echo will provide a clue as to its precise identity. It is evident that among all these sounds of varying frequencies some will prove more helpful than others, to be utilised or discarded according to the needs of the particular situation. Also to be taken into consideration is the fact that the paths of the echoes will vary. Some will be thrown straight back from the object in question, some will be reflected off the bottom or off the surface, and others will have been bouncing back and forth for some time. Here again a selection of the best combination of echoes provides a store of accurate information. The choice is vast for whereas man is normally insensible to frequencies of much over 23,000 cycles per second, dolphins are receptive to frequencies of up to 80,000 cycles.

So keen is the hearing mechanism of dolphins that they can use it for making an accurate distinction between objects of

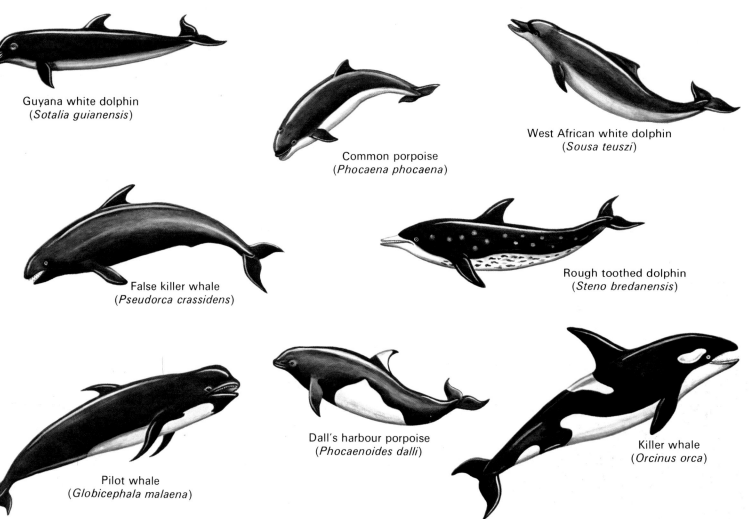

The Delphinidae, largest family of toothed whales (Odontoceti), group together species of widely varying shapes, sizes and colours.

virtually identical shape and outline – a distinction that a man could not even make by visual means. Furthermore they can do this in the case of objects which have similar forms and dimensions but which are composed of different substances. From the point of view of underwater navigation this natural sonar system is well-nigh perfect, as has been demonstrated by a number of experiments with captive dolphins. Set loose in the most complicated mazes, with no possibility of making use of their vision, the animals, once familiar with the test conditions, were able to find their way unerringly through the labyrinth. Apart from proving the sharpness of their hearing, this is clear evidence of their ability to learn.

The mind of the dolphin

Although much research is now being devoted to the subject, we still cannot say anything very positive about the intelligence of dolphins. It is evident, however, from what is already known, that it is very advanced and probably superior to that of the most highly evolved anthropoid apes.

Simple examination of the dolphin's brain is enough to indicate that here is an animal with above-average mental powers. It is relatively larger and heavier than the brain of any primate, including man, and the cerebral cortex (the zone which provides a key to the degree of evolution and intelligence) is of great

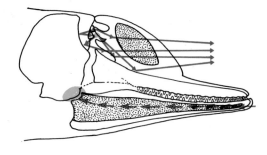

Recent discoveries lead scientists to think that the sounds emitted by dolphins originate in the internal folds of the nasal passages, are reflected from the bones of the skull and upper jaw, and are then amplified by the frontal boss which acts as a soundbox. Hearing is effected through the lower jaw from which the sounds are transmitted to the inner ear.

complexity. Thus the number of convolutions of the cerebral hemispheres is at least double that of the human brain; and there are some 50 per cent more neurons (nerve cells) in the cortex. Apart from inspection of the brain structure, tests with captive dolphins show that they possess a surprisingly high level of intellect. It is interesting to note that all such experiments have only been possible because the animals themselves tend to be so cooperative, taking obvious pleasure in games. This play instinct is in itself regarded as a clue to intelligence. Furthermore, they are quick to learn by imitation, as when sharing a pool with dolphins of other species. They often adopt completely new behaviour patterns quite spontaneously and without expectation of reward.

How dolphins 'talk'

Another fascinating aspect of dolphin intelligence is the manner in which the animals communicate with one another. All zoologists are agreed that the sounds which dolphins emit are not designed solely for purposes of location and detection but also to convey information, though how this is achieved is still a matter of much conjecture. Some authors are convinced that the animals have their own language, almost as elaborate as our own; but although a great deal of work has been attempted, it has not thus far been conspicuously successful, not so much because of errors on the part of the zoologists concerned but because of their inability to find the key which will unlock the system and hopefully lead to significant discoveries.

One thing is obvious, namely that the language of dolphins is very different from ours, consisting not of vocalisations but of frequency modulations. Yet in this connection it is interesting to note that on some of the Canary Islands, where the terrain is mountainous, local shepherds habitually communicate with one another over long distances not by shouting but by means of a complex range of whistles—a system much more akin to that employed by dolphins.

Some zoologists, including Dr. J. C. Lilley, a specialist in this subject, have claimed that it should be possible to teach dolphins to talk like humans, although this is by no means accepted by the majority. It is true that there have been experiments in which tape recordings of sounds made by dolphins, when slowed down, have seemed to reproduce human words, suggesting that the animals have imitated their keepers' commands. Stories of 'talking' dolphins must be treated with caution, but there is one report of a bottle-nosed dolphin which at the start of each training session would murmur (nasally but intelligibly) the phrase, 'Right, let's start!' What cannot be determined is whether such linguistic skill can be taken as proof of intelligence or whether it is simply an automatic response, as in the case of parrots and other 'talking' birds.

Whatever the truth, dolphins are clearly capable of conveying complex information by means of sounds, as has been demonstrated in the interesting experiments of Dr. Jarvis Bastian with a pair of bottle-nosed dolphins. The tests involved the

Dolphins are famous for their graceful leaps high out of the water, powered by remarkably strong tail muscles.

animals in games requiring them to move levers with their snouts in response to light stimuli—the right-hand one if the light was steady, the one on the left if the light was intermittent. When they reacted correctly they were rewarded with a fish. Both soon learned to perform the necessary actions without a mistake. Dr. Bastian then made the game harder. When the light flashed the female had to wait until her partner had pushed the lever before doing it herself; if she went first she received no prize. It did not take them long to learn this variation. Then the two dolphins were separated so that both had their own levers but only the female was able to see the light signal. Furthermore, they could only communicate with each other by echo-location. When the light went on the female waited, as she had been taught, meanwhile emitting various sounds. The male then moved his lever correctly, after which he gave his companion to understand that the moment had come for her to do the same so that both could claim their reward. When separated by a sound-proof barrier the experiment failed; but whenever communication was free the test was successful.

This type of experiment should convince all but the most stubborn observer that some form of dolphin language must exist,

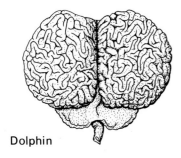

Dolphin

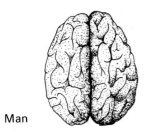

Man

Comparison between the brain of a dolphin and a man provides a guide to the animal's level of intelligence for not only is it larger but also much more complex. There are at least twice as many convolutions of the cerebral hemispheres and about fifty per cent more nerve cells.

even if we cannot begin to comprehend its scope and complexity. Some zoologists will even go so far as to suggest that there is some kind of oral tradition by which information can be transmitted from generation to generation. But our appreciation of dolphin intelligence must take into account the fact that it is very different from our own, notably because we inhabit a visual world whereas dolphins live in a world of sounds. Whatever the nature of their intellect it is quite possible that dolphins may, in the near future, help man in his exploration of the ocean and indeed experiments with this in mind are already afoot. A number of governments have also introduced legislation to prohibit the hunting of dolphins, if for no other reason than that such intelligent animals are especially deserving of protection, indeed killing a dolphin almost amounts to homicide.

The group and the individual

If it is hard enough to study dolphins in captivity it is even more difficult in the wild. Some basic facts have, nevertheless, been established.

Dolphins live in comparatively stable groups and social life appears, in fact, to be absolutely necessary for the mental equilibrium of the individual. Isolated animals (apart from those accidentally separated from their companions) would seem to be exceptions to the rule. Commander Cousteau states that seclusion is literally fatal, the cause of death not being due to any physical defect (such as starvation) but being psychological in origin. Certainly autopsies on such animals have failed to reveal any apparent physical cause.

There does not seem to be any dominant individual leading a school of dolphins; but in captivity a certain hierarchy appears to be established among a number of animals, the dominant invariably being a male.

These social groups certainly act in a cooperative manner, this being best displayed by their habit of coming to one another's assistance, particularly when an animal is in danger. Furthermore, when a female is about to give birth she emits characteristic calls which have the effect of assembling all the other females of the group around her. These mount guard, protecting both the new mother and her baby from the possible attacks of sharks. When the baby is born the mother pushes it towards the surface with her snout so that it can take its first gulp of air; if it comes up against any difficulty she, together with several other females, rush to the rescue, helping it to keep afloat until it can fend for itself. Observations have been made of mothers carrying the carcase of their dead baby around for some days. In the same way, when a member of the school is sick or injured and consequently unable to swim on its own, it will be escorted by two adults who place themselves under the invalid's pectoral fins for support. It is even possible, though not proved, that dolphins practise a form of artificial respiration. The animals normally take a breath every time their blow-hole emerges from the water; and in this particular case the two dolphins supporting their handicapped companion seem to take turns

The common dolphin is a remarkable swimmer which at times reaches a speed of around thirty miles per hour.

in pushing it up to the surface so as to keep it breathing. Releasing their hold alternately, they too are able to come to the surface to breathe. Other observations indicate that various adults in the school take turns to assist a disabled companion and that the latter may even be helped by dolphins of other species.

Schools containing more than one species are frequently seen – further proof of their sociability. Undoubtedly some form of communication exists between dolphins of the genera *Tursiops* and *Globicephala, Lagenorhynchus* and *Globicephala,* and *Lagenorhynchus* and *Delphinus*. Other unexpected associations in captivity involve dolphins and Californian grey whales, and even dolphins and killer whales, despite the fact that in the wild the latter prey freely on their smaller relatives.

Dolphins around the world

The term 'dolphin' is used to describe a large number of cetaceans belonging to the suborder Odontoceti (toothed whales). Classification is not simple. Apart from the members of the family Delphinidae (which includes certain whales as well as

Two well contrasted representatives of the family Delphinidae are the spotted dolphin (*above*) and the pilot whale (*below*).

dolphins) there are freshwater dolphins of the family Platanistidae and large dolphins of the family Ziphiidae, the largest of which (*Berardius*) may be up to 40 feet long. Altogether there are approximately 60 species.

Broadly speaking, the toothed whales can be divided into three major groups—the cachalots or sperm whales (Physeteridae), porpoises (Phocaenidae) and dolphins (Delphinidae, Ziphiidae, Stenidae and Platanistidae). Apart from these are the beluga or white whale and the narwhal, both of which belong to the family Monodontidae.

According to the classification of Dale W. Rice, the family Physeteridae comprises two genera, *Physeter* and *Kogia*; and the Phocaenidae contain three genera, *Neomerus*, *Phocaena* and *Phocaenoides*. The Delphinidae are made up of fourteen genera, namely *Feresa*, *Globicephala*, *Orcaella*, *Orcinus*, *Peponocephala*, *Pseudorca*, *Lissodelphis*, *Cephalorhynchus*, *Delphinus*, *Grampus*, *Lagenodelphis*, *Lagenorhynchus*, *Stenella* and *Tursiops*. The five genera of Ziphiidae are *Berardius*, *Mesoplodon*, *Hyperoodon*, *Tasmacetus* and *Ziphius*. The family Stenidae groups together *Sotalia*, *Sousa* and *Steno*. The four freshwater genera of the family

Platanistidae are *Platanista, Stenodelphis, Inia* and *Lipotes*; and the two genera comprising the Monodontidae are *Delphinapterus* and *Monodon*.

The representatives of the genus *Steno* (included among the long-snouted dolphins) are found in the tropical waters of the Atlantic, Pacific and Indian Oceans as well as in the Gulf of Bengal, the Red Sea, the Mediterranean and the Caribbean, but their habits are a complete mystery. Dolphins of the genus *Sousa* are inhabitants of the warm regions, both salt and freshwater, of southern Asia and Africa, and resemble the bottle-nosed dolphins in appearance. The small dolphins of the genus *Sotalia* prefer freshwater zones, with species along the coasts of Guyana and Brazil and throughout the Amazon basin.

Turning to the Delphinidae, we find the many species of the genus *Stenella*, some of them distinctively spotted, distributed through all the warm oceans of the world. The dolphins of the genus *Delphinus*, including the common dolphins frequently depicted in Greek frescoes, have an equally wide range but also frequent temperate seas. Both genera have a grooved beak.

Risso's dolphin (*Grampus griseus*) has recently made appearances off the east coast of the United States, in the North Atlantic, off South Africa, in the Mediterranean and Red Sea, and off the shores of China, Japan, Australia and New Zealand. This grey animal measures about 13 feet long and lacks a beak. Pelorus Jack, the celebrated dolphin protected by the New Zealand government, belonged to this sociable species.

The bottle-nosed dolphin (*Tursiops truncatus*), much used for experimental study and for entertainment purposes, has a world-wide distribution. Rather less playful but extraordinarily sociable, mixing freely with other species, are the short-beaked dolphins of the genus *Lagenorhynchus*.

The habits of the two species of pygmy killer whales of the genus *Feresa* are little known but they are believed to have a global range. They look somewhat like the pilot or caa'ing whales of the genus *Globicephala*, which inhabit all oceans apart from polar seas, but are considerably smaller.

The oceans of the southern hemisphere, especially cold waters, are the haunts of the little known Commerson's dolphin (*Cephalorhynchus commersonii*). It is also called the piebald dolphin because of its black and white coloration. Sporting the same spectacular colours but far more dangerous is the dreaded killer whale or grampus (*Orcinus orca*), most terrible of all sea predators, which is found almost everywhere but especially in the Arctic and Antarctic Oceans. Slightly smaller is the false killer whale (*Pseudorca crassidens*).

The Irrawaddy dolphin (*Orcaella brevirostris*) frequents the coasts of South-east Asia and has been sighted up to 900 miles from the mouth of the river of that name.

Perhaps the most curious members of the family are the two right whale dolphins of the genus *Lissodelphis*, both of which lack a dorsal fin.

So named because it appears to be an intermediate form between *Lagenorhynchus* and *Delphinus* is *Lagenodelphis hosei*, known only from a skeleton from the River Lutong in Borneo.

When a sick or injured dolphin is unable to swim on its own two of its companions may come to the rescue, placing themselves below the body of the handicapped animal. They push it towards the surface so that it can continue to breathe and take it in turns to surface so that their own breathing is not interrupted.

Facing page: To watch a school of dolphins cutting through the water with consummate ease and grace is to realise that these animals are among nature's evolutionary masterpieces.

The porpoises of the family Phocaenidae (actually small beakless dolphins) are well known because of their large numbers and frequent presence off coasts and in rivers where they often breed. The common porpoise (*Phocaena phocaena*) is the most familiar European species, black above and white below. The Pacific porpoises of the genus *Phocaenoides* apparently undertake seasonal migrations. The finless black porpoise (*Neomeris phocaenoides*) is an Asiatic species, found off coasts and in estuaries, rivers and lakes, having been sighted 1,000 miles up-river from the mouth of the Yangtse Kiang. It is very small (about $4\frac{1}{2}$ feet) and lacks a dorsal fin.

The common dolphin

The common dolphin (*Delphinus delphis*) has captivated the fancy of man since ancient times. Portrayed on mosaics and frescoes, mentioned in numerous myths and fables, this friendly animal, the most playful of all dolphins, has been credited with saving the lives of many a shipwrecked sailor by guiding them to shore. Because of such tendencies one might assume that the species would be an ideal subject for scientific research but this is not the case for it does not do well in captivity. Accustomed to life on the high seas, the common dolphin simply does not adapt well to confined spaces.

The common dolphin possesses a larger number of teeth than any other land or sea mammal – 45-50 pairs in either jaw. Feeding principally on animals living on the sea bed, the species has almost no natural enemies apart from the killer whale; and since it is an excellent swimmer that can reach speeds of 30 miles per hour and more, sheer pace often enables it to escape these fiercest of all predators.

It has been possible, nevertheless, to study two reflex actions of this species connected with breathing, which appear to be common to all dolphins. The first involves breathing in and out when the blow-hole is above the water surface; the other is a powerful, automatic up-and-down movement of the tail. In combination they ensure that respiration only occurs when the spiracle is exposed, thus diminishing the risk of the animal breathing in any water.

Reproductive groups sighted in the Black Sea have consisted of six to eight males and one female. Mating takes place either on the surface or underwater. Gestation apparently lasts ten to eleven months and the majority of births occur throughout the winter or in midsummer.

The friendly bottle-nosed dolphin

The docile and friendly bottle-nosed dolphin (*Tursiops truncatus*) is the one which is most often in the public eye – central attraction of dolphinariums and star of cinema and television screens. So named because of its well-defined beak, it feeds principally on fishes (including shoal-forming pelagic species) as well as on invertebrates such as bivalve molluscs. According to Dr. F. C. Fraser, it also attacks small sharks.

COMMON DOLPHIN
(*Delphinus delphis*)

Class: Mammalia
Order: Cetacea
Family: Delphinidae
Length: 59–102 inches (150–260 cm)
Length of pectoral fin: 12 inches (30 cm)
Height of dorsal fin: 24 inches (60 cm)
Breadth of caudal fin: 20 inches (50 cm)
Weight: 165 lb (75 kg) or more
Diet basically fishes and cephalopods
Gestation: 10–11 months
Number of young: one

Slender, streamlined body. Colour variable; back dark brown or black, belly white; wavy grey or beige stripes on flanks. Dark stripes from eyes to muzzle. Forty-five to fifty teeth in each half-jaw. Narrow beak.

Some dolphins adapt so well to life in captivity that scientists have been able to carry out searching experiments on their biology and behaviour. They get on well with humans and often form surprising associations with other species including, as shown in this photograph, Californian grey whales.

Bottle-nosed dolphins live in groups consisting of individuals of both sexes and all ages, their number varying according to availability of food. Although there appears to be no single leader of a school, there is a recognisable hierarchy based on size. In the open sea the animals often swim some hundred yards apart from one another but nearer to shore they close ranks to a distance of 30–50 feet. Members of the same school recognise one another and stay together by emitting sounds of varying register; and for some reason yet unexplained there is no confusion among groups living in the same area.

Normally sedentary by habit, bottle-nosed dolphins adjust well to life in captivity and are extremely cooperative and understanding test subjects. Most experiments with dolphins have been with this species. They seem to be able to distinguish between the types of problems set for them, showing themselves very willing and docile when these are simple but abandoning any that are too difficult, no matter what the rewards offered. They are undoubtedly highly intelligent but sometimes they display behaviour that can only be described as neurotic, apparently arising from their captive state. These fits of depression can only be relieved by sedative drugs.

Bottle-nosed dolphins are active by day and rest at night, but their period of sleep is very brief, being frequently interrupted in order to breathe. Every meal is followed by a short and similarly discontinous siesta period. Males do not seem to sleep in the same manner as females for they float with their blow-hole out of the water whereas their companions remain a foot or so under the surface. Both sexes are capable of diving down to 60–70 feet, staying below for fifteen minutes or more. The rate of heartbeat is variable, according to the circumstances, ranging from 108 per minute at the surface to a mere 50 or thereabouts under water. These fluctuations are doubtless due to adaptation which encourages them to conserve oxygen when submerged.

The breeding season lasts from spring to summer and the gestation period is 10–11 months. The babies are born tail-first, like all cetaceans, and are suckled for a variable time, anything from six to fifteen months. Feeding of the young takes place in the water, such sessions consisting of from one to nine sucking attempts only, each of which lasts a few seconds, for the muscles of the mammary glands drive out the milk under high pressure. The animals reach sexual maturity at the age of five or six years.

The terrifying killer whale

The grampus or killer whale (*Orcinus orca*), most commonly found in Arctic and Antarctic waters (although represented in most other oceans because of its omnivorous tastes and tolerance of a wide range of temperatures), is not only the most ferocious hunter of the seas but also the fiercest predator on earth. The species does not appear to have migratory habits. An enormous range of food includes gregarious fishes, cephalopods, sea mammals and sea birds.

Males of this species measure up to 30 feet in length but females average only 13–20 feet. Yet even a small female may weigh more than three-quarters of a ton.

The jaws of these colossal killers are lined with powerful, conical teeth, from ten to fifteen in each half-jaw. The stamina, strength and athletic prowess of the animals are astonishing. Although their cruising speed ranges from 6 to 8 miles per hour, top speed is believed to be nearer 22–23 miles per hour. Furthermore, they are capable of leaping clear out of the water to a height of 3–4 feet and over a distance of more than 40 feet.

Killer whales live and hunt in packs made up of anything from three to fifty individuals. For sheer aggressiveness, yet also for courage, they are unrivalled in the animal kingdom. They frequently pursue cetaceans larger than themselves, showing no fear of any species. It is an established fact that their victims, no matter of what size, are frozen into immobility in their presence. Although such behaviour has been interpreted as being the result of intense panic which paralyses flight reactions, the truth is probably far simpler. An animal which remains immobile has a much greater chance of escaping the sonar locating system.

BOTTLE-NOSED DOLPHIN
(*Tursiops truncatus*)

Class: Mammalia
Order: Cetacea
Family: Delphinidae
Total length: 69–142 inches (175–360 cm)
Length of pectoral fin: 12–20 inches (30–50 cm)
Height of dorsal fin: 12 inches (30 cm)
Breadth of caudal fin: 24 inches (60 cm)
Weight: 330–440 lb (150–200 kg)
Diet: basically benthic fishes
Gestation: 10–11 months
Number of young: one

Strong but streamlined body. Beak short, lower jaw longer than upper one; twenty to twenty-six teeth in each half-jaw. Upper parts greyish-brown or black with violet tints; underparts light grey or white. At birth baby measures about 3 feet and weighs 25 lb.

KILLER WHALE OR GRAMPUS
(*Orcinus orca*)

Class: Mammalia
Order: Cetacea
Family: Delphinidae
Total length: male 20–30 feet (6–9 m)
female 13–20 feet (4–6 m)
Length of pectoral fin: up to one-sixth total length
Weight: upwards of 1,870 lb (850 kg)
Diet: carnivorous
Gestation: 10–11 months
Number of young: one

Very large, heavy body, shining black above, white or yellowish below, with other white marks behind eye and about half-way between pectoral and caudal fins. Muscular tail and very large dorsal and pectoral fins, the former straight and vertical, forming an almost perfect isosceles triangle over 3 feet tall; pectoral fins large and rounded. Flat snout, enormous mouth and throat; ten to fifteen powerful, conical teeth in each half-jaw; at birth baby measures about 6–8 feet.

The incredible power and ferocity of the killer whale are greatly in evidence in the special hunting technique often adopted to capture seals in Antarctic waters. The fact that the seals may be basking on pack-ice is no guarantee of their safety, for the terrible predator of the deep will dive down and then shoot straight up, smashing the ice with its massive snout and tumbling the seals into the water, where they are easy prey.

One of the most spectacular examples of the aggressive hunting techniques of killer whales occurs in Antarctic seas. Having located a group of seals basking on the pack-ice, the whales may dive down and rocket up through the ice, even when it is several feet thick, to catch their prey unawares. If by some error of calculation the first attempt proves unsuccessful they will repeat the operation for as long as may be necessary to tumble their victims into the water as the ice cracks. Once this happens the seals are doomed.

The false killer whale may be distinguished from the true killer whale by its uniform colour (no white underparts), its more swollen snout and its shorter dorsal fin and smaller flippers.

Surprising as it may seem, killer whales occasionally travel far up rivers and a number of sightings have been reported from different countries. Thus in October 1931 a female grampus, measuring about 12 feet long, entered the Columbia River in Oregon and was seen at intervals over a month feeding on freshwater fishes. Another female killer whale, almost as large, swam twenty miles up the Firth of Forth in Scotland, following salmon. Ten killer whales have been killed in the River Parrett in Somerset and three in the Thames. Two large adult whales which entered a small bay at Eastport (Maine) in March 1902 remained in the river for about four weeks. Like other dolphins, killer whales often follow in the wake of ships, either out of curiosity or a tendency towards playfulness.

The gestation period of killer whales is comparatively long. According to certain authors it lasts sixteen months; but this is probably an exaggeration and 10–11 months may be nearer the truth. Births generally take place in November or December, babies measuring 6–8 feet.

Despite their huge appetite and ferocious tendencies, killer whales evidently do not attack humans; and the few individuals that have been successfully reared in captivity appear to be surprisingly docile, playing with their keepers and even allowing them to ride astride their back. Observations of such whales in captivity suggest that their mental ability is on a par with that of the more familiar dolphins. Thus there is some justification in regarding them not only as the strongest but also as the most intelligent of the world's predators.

The magnificent killer whale, with its enormous jaws and throat, is the most fearsome predator on earth. Males may measure up to thirty feet. In the ocean it plays the role of super-predator, hunting in groups and claiming a wide range of victims, including baleen whales.

CHAPTER 12

The wide oceans and the abyss beneath

Of the three major marine zones—the seacoast, the continental shelf and the high sea—the last-named is by far the most extensive. The immense expanses of open ocean, by virtue of the fact that sea water is more or less unvarying in composition no matter where it is found, are more homogeneous and continuous than the other two regions. But although their main characteristics are alike, wherever they are situated, there are obviously certain regional differences. There is a vertical distribution of marine organisms depending on the depth of the ocean; and there is a horizontal distribution determined by geographical location. The result is that the fauna and flora of the open sea tend to be separated into distinct marine zones, but that these are not so sharply isolated from one another as are the animals and plants of the zoogeographical regions existing on land.

Evaporation reaches its peak in warm waters and this engenders a stronger concentration of sodium chloride (salt), increasing the water density. The same phenomenon, but for other reasons, occurs in circumpolar seas where the very low outside temperature leads to the formation of immense slabs of ice (mainly composed of fresh water) so that the sea water which remains unfrozen is also of high density.

Starting at the South Pole (where the volume of water is far greater than that of the land) and progressing towards the North Pole, it is possible to divide the oceans into a series of zones, each of which contains distinctive fishes, mammals and birds. In the far south are the cold waters surrounding Antarctica, which are commonly divided into two zones, the antarctic and subantarctic, separated fairly sharply by the so-called line of antarctic convergence where there is an abrupt change of

Facing page: Recent exploration of the world under the sea has brought to light many astounding discoveries, such as that of the coelacanth, sole representative of a group of fishes whose fossil remains date back two hundred million years.

temperature. To the south the seas are colder, to the north they are warmer, the difference in temperature being estimated at more than 5°C.

Travelling northwards, there is a noticeable, though less sharply defined, change around latitude 42°S where subantarctic waters meet warmer subtropical waters. In this southern subtropical zone the water temperature may reach 20°C and the salt content is higher; at the same time there is a gradual decrease in the content of dissolved gases and the concentration of nitrates and phosphates. Conditions are therefore less favourable for the development of plankton and as a result there are rather fewer marine animals.

In the tropical zone, girdling the equator, the water temperature reaches 30°C, salinity is even higher and plankton scarce, But the general structure of this zone is influenced by the Humboldt, Falkland and south equatorial currents which modify the water temperature to some extent and permit animals to pass from one latitude to another.

The bizarre, sometimes monstrous, shapes of the fishes of the abyssal depths, are adaptations to a life spent in pitch darkness and under high pressure. The species shown here, from left to right and top to bottom are *Chiasmodus niger, Lamprotoxus flagellibarba, Argyropelecus affinis, Photostomias guernei, Chauliodus sloanei, Diretmus argenteus, Platyberyx opalescens* and *Melanocetus johnsoni.*

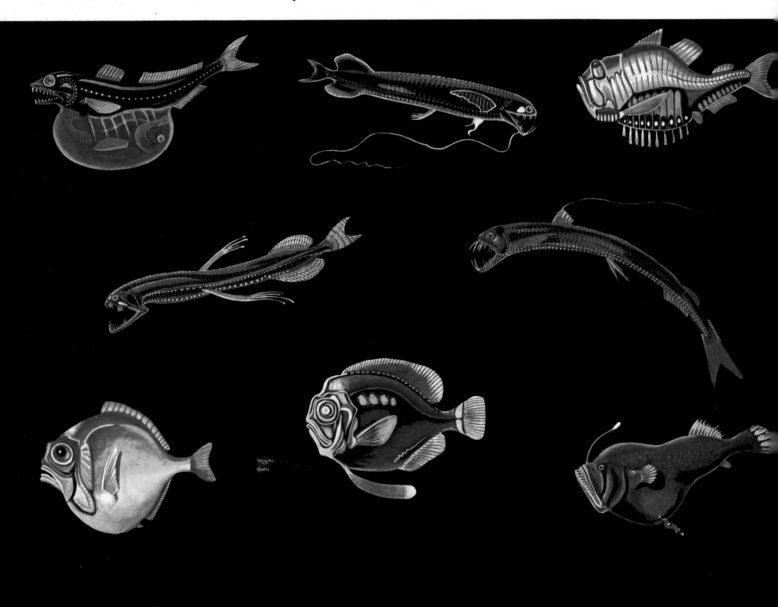

In the northern hemisphere there are three zones – northern subtropical or temperate, boreal and arctic. Here the distribution of fauna and flora is less complex for the course of the various ocean currents (Gulf Stream, Labrador, Kuroshiwo and California, to mention just the main ones) are diverted by continental land masses so as to bring about irregularities of temperature. Such fluctuations modify the theoretically concentric pattern of the different ocean zones and are the reasons why the seas of the northern hemisphere, with lower temperatures and salinity, support a much more varied and thriving range of animal and plant communities.

The upside-down mountain

Only recently has progress in technology made it possible for man to begin exploring the silent world under the sea. Until then his conception of this mysterious watery realm was based on nothing more than legend, imagination and superstition. Now he has come face to face with reality, setting eyes for the first time on the most secret places of the planet, venturing down to depths so far removed from the surface that no ray of sunlight illuminates them.

If in the course of an underwater journey a diver were to move gradually outwards from the coast to the deepest waters of the abyss, he would see that the marine plants and animals are arranged vertically, similar to the distribution on a mountain, but in reverse. Beyond the continental shelf, at about 600 feet, the ground slopes more sharply, sometimes plunging straight down, and this is the frontier of the deep sea. If our man were particularly observant he would note subtle changes as he sank from the slope to the abyssal depths, including the appearance of species better and better adapted to the submarine environment, as well as a decline in the actual number of species with increasing depth. Both phenomena have analogies with living forms on a mountainside where the degree of adaptation and diminishing number of species are directly related to altitude. On dry land great height entails shortage of oxygen, lowering of pressure and falling temperature. In the ocean depths a parallel process leads to a gradual reduction of light and warmth, an increase of pressure and a marked decline in the supply of oxygen. Thus it is not being fanciful to compare the underwater world to a gigantic upside-down mountain or to a huge inverted mould, inside which, attached to the walls or suspended in a mass of water, is a complex living community whose nature is determined by the great depth at which they live and the unusual characteristics of their environment.

The abyssal depths

The continental slope joins the abyssal plain at depths ranging from 12,000 to 17,000 feet and beyond it stretches the true ocean depths, above the abyssal plain. First descriptions of this region, based on flimsy data, compared it to an immense basin with completely smooth edges. But soundings taken with

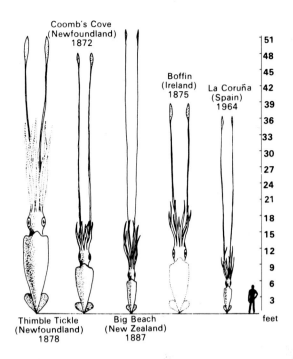

The giant squids of the genus *Architeuthis* are very probably responsible for many of the sailors' legends about sea monsters. Here, compared with an average-sized man, are five such squids which have actually been found and measured within the last century or so in various parts of the world.

The skeletal remains of countless organisms have accumulated on the ocean floor and helped to make up the characteristic soft, muddy substratum. Among them are the siliceous shells of diatoms and other microscopic algae of varied shapes, one algae of which is shown in this photograph.

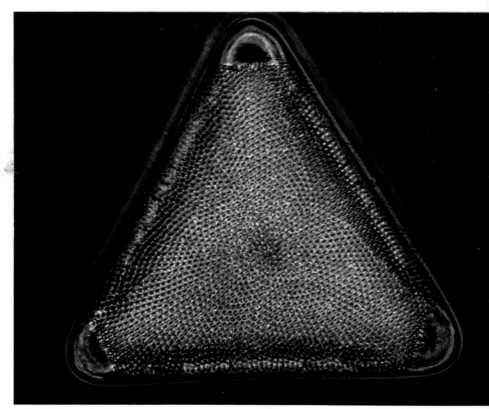

Medusas are typical inhabitants both of the shallow waters of the continental shelf and the high seas.

electronic instruments similar to radar, reinforced by direct observation, have now established that the submarine depths, far from being uniformly flat, are in many ways similar in form and structure to the surfaces of the dry land masses. The abyssal plain is cross-crossed by mountain chains, with plateaux, gorges and trenches. In the areas lying closest to the continents, the furrows of underwater rivers can be seen. The relief of the continental slope itself, composed as it is of materials originating on land, is made up of channels and ridges formed by the continuous flowing action of water. Although considerably more vast, they are comparable to the artificial trenches dug on land to accommodate roads and railway lines.

The materials which geographical erosion has ripped away from the continents have gone to form the contours of the ocean bed. In places the abyssal plain is carpeted with a thick layer of mud or extremely fine sand, such deposits having accumulated as a result of the dead-calm conditions in the depths. The slightest disturbance would disperse the particles (whose diameter is as a rule less than one-fiftieth of a millimetre) and leave them floating in the water. But most of the sediment is of organogenic origin, produced by living creatures, including countless millions of diatoms and protozoans (Radiolaria and Foraminifera) whose unicellular bodies are covered by a solid siliceous or calcareous shell and which are forever reproducing and dying. The remains of these organisms fall like a continuous rain onto the bottom and there they accumulate. Here too are the skeletons of larger animals, decomposed as a result of bacterial action; and the general panorama of materials making up the complex structure of the ocean floor and providing food and shelter for innumerable benthic species, is completed by substances thrown up by periodic volcanic eruptions.

The most obvious change in conditions as the water depth increases is the gradual loss of light. In fact, beyond the 1,350-foot mark it vanishes entirely. At this depth and in the darkness below there is no further possibility of green plant growth and consequently there are far fewer animals. The rare phytophages have the utmost difficulty in procuring sufficient food at these levels, for the only nutritious substances are the residues of dead plants which fall gently and continuously down from the surface to the ocean floor.

There is a parallel decline in the oxygen content of the sea water. Although in the intermediate zones between the surface and the bottom there is still enough of the gas in dissolved form to support life, in some abyssal trenches it is completely absent; in such areas the only surviving organisms are a few anaerobic bacteria.

The temperature in these submarine depths is also very low, never exceeding 4°C, and, in the zones farthest removed from the surface, hovering around the 0°C level. As for pressure, this is considerably increased, at a rate of about one atmosphere for every 30 feet, suggesting that in the deepest parts of the ocean it is equivalent to some 1,100 atmospheres.

It is obvious, therefore, that only a small but highly adapted living community can withstand the conditions of the abyss.

One of the many strange and spectacular fishes of the abyssal depths is *Eretmophorus kleinenbergi*, notable for its extraordinarily long pectoral and ventral fins, with feather-like tips.

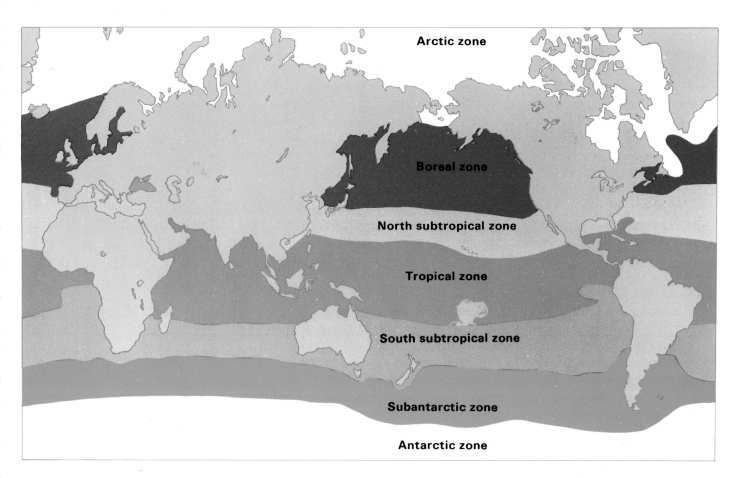

The ocean, like the land, is geographically divided into zones, according to the temperature and salinity of the water. The lines separating the different zones assume an undulating pattern because they tend to follow the courses of warm or cold ocean currents.

The curious animals of the abyss

Exploration of the immense abyssal depths is a comparatively recent science. Indeed the information thus far accumulated is so fragmentary that the most that can be said is that man has only just begun to probe the mysteries of the dark ocean depths; and it will be some time before anything like a comprehensive picture emerges. The first examinations of the true abyssal regions (at depths of more than 20,000 feet) date only from 1948, so that allowance must be made for the fact that this is a very young branch of oceanography and that its methods are still limited. Nets and dredges are not only extremely difficult to manoeuvre when suspended at great depths from the ends of cables several miles long, but also have to be handled with the utmost care when being hauled up. Furthermore, the samples so far brought to the surface are an inadequate representation of the abyssal fauna in its entirety, given that the animal inhabitants of the ocean depths are few in number and often so large and fast-moving that it is virtually impossible to catch them.

Nevertheless, in spite of all difficulties, enough data are already available for oceanographers to trace, with some measure of accuracy, a structural outline of the organisation and development of life in the ocean. At depths of more than 600 feet there are many species adapted to an enviroment where the substratum consists of mud and ooze. Most of these species possess some form of stalk so that they are held up above on the mud particles; they include sea anemones, sea lilies and sea pens. Here too are a number of crustaceans, well equipped to move freely over the muddy bottom, their long, slender legs providing a firm grip on the soft, slippery surface. Examples are the prawns of the genus *Nematocarcinus*.

Other groups burrow into the ooze of the ocean floor, such as the molluscs of the genera *Dentalium* and *Cadulus*, known as tusk shells, various sea urchins (Cidaridae) and tube worms which find refuge in the grains of sediment formed by the skeletal remains of other marine animals. Here too, anchored in the mud, are diverse sponges (Hexactinellida), several coelenterates and, where the substratum is firm, cirriped crustaceans including those of the genera *Scalpellum* and *Verruca*.

After the initial exploration of the abyss, in the course of which a surprisingly large number of invertebrates and fishes were recovered (and these very different from the familiar species already encountered in coastal waters), scientists speculated that the ocean depths might harbour primitive forms of animal life similar to fossils discovered on land. But the various oceanographic expeditions that have since been undertaken have shown that the most primitive marine species are to be found, not in the abyssal depths, but in the bathyal zone of the ocean at depths of between 650 and 3,500 feet. When in 1938 the first coelacanth was brought to the surface, off the south-east coast of Africa, it was assumed that this fish (thought to have been extinct for 70 million years) had emerged from the shadows of the deep and that only a favourable combination of circumstances had permitted it to survive in comparatively shallow waters. But the discovery of other coelacanths subsequently made it clear that these fishes are for the most part inhabitants of regions above the abyss, living alongside species of much more recent origin. This fact leads many zoologists to believe that the animal communities of the abyss may be the most modern of all the ocean populations and that their development and evolution are as yet incomplete.

This is not to say that all the animals of the abyssal plain are necessarily of recent origin, for the *Galathea* expedition came up with some extremely primitive forms. Thus one dredge carried out off the Pacific coast of Mexico, at a depth of more than 10,000 feet, brought to light certain molluscs which bore a superficial resemblance to limpets. These, however (which were subsequently given the name of *Neopilina galathea*), proved to be the last surviving representatives of the Monoplacophora, a group thought to have been extinct since the Paleocene, at least 400 million years ago.

Life in the abyssal regions is characterised by certain peculiarities which distinguish this environment completely from any other marine habitats. In the first place, as a result of the perpetual darkness, there are no green plants and hence no true plant-eating animals. The ecological pyramid or food chain is thus represented here by two basic groups only, detritivores and predators. The former, as their name suggests, feed exclusively on dead organic substances (detritus) which drop from the upper layers of the ocean and pile up on the ocean bed. Consumers of such materials include, for example, sea cucumbers — echinoderms with a long, sausage-shaped body which is almost entirely composed of water and provided with foot-like tentacles

Posed specimens of the silver hatchet fishes of the genus *Argyropelecus*, recognisable by their protruding eyes and luminous organs on head and flanks. These fish are widely distributed in the abyssal regions.

The molluscs which have adapted best to life on the high seas are the cephalopods. These include the common squids of the genus *Loligo*, shown here, which gather in shoals, as well as giant squids inhabiting the abyss.

for moving about and scooping up food on the muddy bottom. Several brittlestars, such as those of the genus *Amphophiura*, and sea spiders likewise feed on detritus, most of them simply swallowing the mud and extracting what little nutritive substance they can, a process similar to that used on land by earthworms. And, of course, there are marine bristleworms, related to earthworms, which feed in the same way.

The majority of these detritivores store enormous quantities of water in their bodies. Consequently the metabolic rate of such individuals is extremely low. They require only a very little food and they live for a long time. The weight of dry matter is never more than 10 per cent of the total body weight, and in the case of sea cucumbers seldom exceeds 3 per cent. The result of this is that the animals preying on them obtain an infinitesimal quantity of protein and this is a serious obstacle to their development and chances of survival. Indeed, observations carried out by a number of oceanographical expeditions indicate that at a depth of about 16,500 feet there exists no more than 5 grammes of dry matter per square metre, a very low quantity which at first glance would seem wholly insufficient to feed a community of predators. How an acceptable balance between predators and prey is in fact maintained is one of the many problems relating to life in the abyss which scientists still have to resolve.

Apart from these predators living on the ocean floor, notably the starfishes, there are many hunting species of molluscs, crustaceans and fishes that move about in these deep waters, following a life pattern which has entailed the acquisition of a variety of adaptations. As already noted, many abyssal species are capable of giving out light by means of luminous mechanisms not all that different from those of terrestrial forms such as glowworms. This involves the production of a type of luminous energy – cold light – which entails no important loss of calories, emitted by special organs known as photophores, in which a biochemical oxidation process occurs. The luminous substance is the outcome of an interaction between the compound luciferin and the enzyme luciferase. The majority of abyssal species are furnished with photophores and it would be impossible within this brief space to describe all the many physical transformations which are involved, according to the hunting patterns of the animals concerned. But the luminous apparatus most frequently in evidence is located at the tip of a long filament on the head which serves as a kind of fishing rod, the light acting as a lure for small fishes and crustaceans. On the other hand, many species make use of this luminous apparatus for protection against enemies. Thus certain crustaceans spill out clouds of light (similar to the ink jets exuded by octopuses and squids) in order to mislead their predators.

The ocean floor is the last layer of the inverted mountain whose base is at the surface. From the zoological point of view, the photic zone, less than 600 feet deep, is the most productive part of the ocean. Below this lives another community, dependent on that of the photic zone, which makes periodic migrations to the shallower levels in search of food. A little deeper there is another compact community which is sustained by the one immediately above; and lower down, on the solid substratum, living organisms feed on the dead and waste substances dropping from the upper layers whilst themselves being preyed upon by hunters at the same level and from above. The scarcity of food at these lower levels and its abundance close to the surface are the main reasons for vastly differing conditions of life, competition being much more intense as the depth increases and the animal community correspondingly more sparse.

The legendary giant octopuses

Science still knows comparatively little about the inhabitants of the deep sea which are not fixed to the bottom. Some of them are very large and extraordinarily mobile, so that not many have been caught in drag-nets. But the few that have come to light are so unusual in shape, with so little superficial resemblance to familiar inshore species, that it is hardly surprising they should have given rise to a wealth of legends. The history of the sea abounds with stories of sinister monsters, of enormous whales that have overturned the flimsy craft of ancient navigators (some of the animals being so large as to have been mistaken for islands)

The ocean floor, usually formed of mud or sand or ooze, is the home of many animals which depend for their food on substances floating down from higher levels. This holds true both for fishes and echinoderms such as sea urchins, pictured here.

and of gigantic octopuses whose huge tentacles have crushed boats and their entire crews. Such tales testify to the power of human imagination and primordial fears of the unknown even in otherwise hard-headed people.

Sailors seem to have been particularly obsessed with the conception of the giant octopuses, for stories featuring such animals were being fabricated long before science revealed that large cephalopods actually did exist. Lurid and highly fanciful descriptions of first-hand encounters with these monsters were supported by masters of whalers who swore that they had caught sperm whales, inside whose stomachs were found entire and still intact bodies of immense octopuses.

Although close examination revealed most of these tales to be pure invention, science has since proved that gigantic cephalopods — admittedly squids and not octopuses — do roam the seas. The capture of astonishingly large specimens not only confirms some of the myths but indicates that these species (belonging

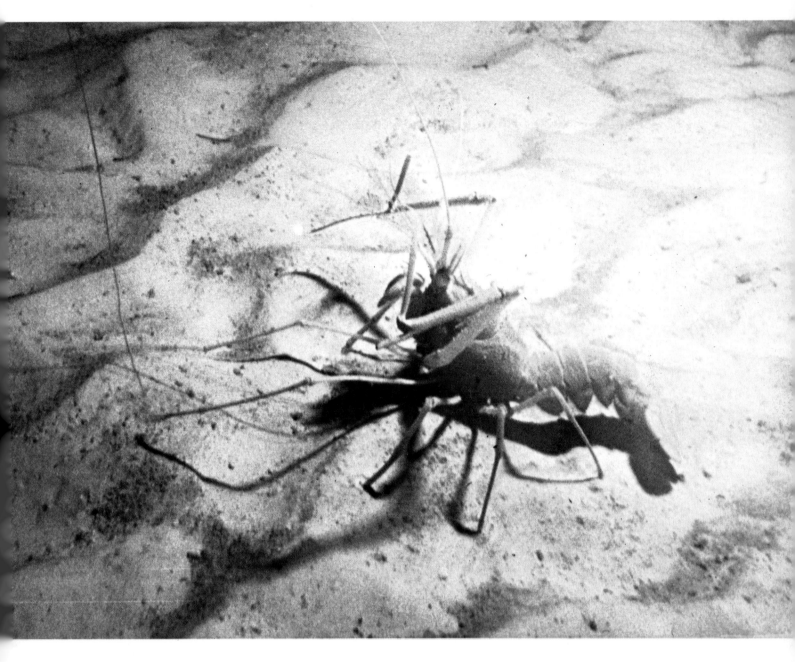

to the genus *Architeuthis*) are genuine inhabitants of the abyss which travel up to shallower zones for food. The size of these animals, including the body and the full length of the tentacles, is usually more than 30 feet and has in some cases exceeded 100 feet! The soft parts of the body, thick and muscular, are so structured as to facilitate upward and downward movements through water without being affected by changes of pressure. The dimensions and predatory habits of these squids makes them veritable lords of the deep and one can imagine that fights between such creatures and powerful sperm whales must be spectacular affairs indeed. Even if whole bodies have not been discovered, the presence of massive tentacle fragments in the mouths of the whales as well as deep gashes from the squid's suckers on their skin bear witness to the bitter struggles that must sometimes take place between these titans of the sea. Clearly we must think twice before dismissing all the ancient legends of the seas as mere fiction.

The animals of the abyss – a region characterised by absence of light and hence of green plants – are broadly divided into detritivores and predators. The latter group includes numerous species of crustaceans.

CHAPTER 13

Killers of the deep

Although few people will ever have the terrifying experience of coming face to face with a shark in the open sea, the mere sight of that unmistakable silhouette in a photograph is often enough to send shivers up the spine. For the shark is as legendary a killer as the lion, the tiger and the leopard; and its fascination is almost wholly derived from the fact that, like them, it is reputed to be unusually ruthless.

In a sense, however, the shark is an even greater mystery than the powerful carnivores of the jungle and bush. After all, they have been hunted for centuries, with the result that the survivors have either been driven into the remotest parts of the wild or have been rounded up in reserves as a harmless tourist attraction. Man can congratulate himself that he has gained the upper hand. But the shark has never been persecuted in the same way. Inhabitant of the vast oceans to which man is a stranger, it ranges freely over about half the earth's surface, its sinister shape symbolising latent menace. Little wonder that man, who has never begun to master it, feels distinctly uneasy in its presence, and this is made worse because the behaviour of sharks is especially unpredictable.

The majority of sharks are inhabitants of tropical and subtropical seas but in fact individuals may be met in all the world's oceans apart from the polar seas. Some species have even adapted to fresh water (as in Lake Nicaragua in Central America) and have been sighted more than 40 miles from the mouths of such large rivers as the Zambezi, the Ganges and the Amazon.

Despite their abundance and wide range, little is known about the behaviour of sharks; and it is precisely because of the absence of verified facts that rumour and legend have found such ready acceptance. Although many such tales clearly defy all logic, sharks seem to have been universally condemned as ferocious

Facing page: Because their domain has only recently begun to be explored by man, the great predators of the oceans, such as the sharks, have not been hunted so relentlessly as their counterparts on land. The various species of sharks still range over about half of the earth's surface.

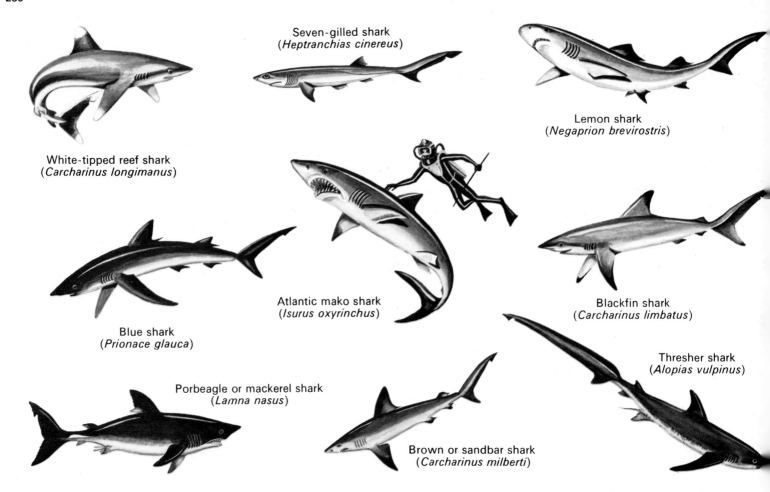

killers with insatiable appetites, prepared to attack everything in range. Sailors have long tended to look on them with superstitious dread as the very symbol of death; but this blind fear is now being proved to be based on little more than ignorance. Zoologists who have studied their behaviour, whilst admitting that some species are indeed very dangerous, point out that others are as harmless as dolphins.

The predatory sharks

The first sharks appeared in the oceans approximately 350 million years ago. Since then species have evolved and disappeared, but basically modern sharks differ very little from their remote ancestors. This continuity of development indicates how efficiently they are adapted to their role of predators extraordinary.

The streamlined shape of the shark—cutwater snout, slightly flattened body, pointed fins and large, upturned tail—is designed for rapid movement through the water. This great swimming capacity is in fact necessary for survival; unlike other fishes, the shark does not have a swimbladder which enables other fishes to float up and down at will and remain motionless at a suitable depth. Lacking this organ, a shark must be continually on the move.

Even if a shark did possess a swimbladder it would still need to keep swimming in order not to die of asphyxia. Because of the position and lack of mobility of its gill openings, the animal is compelled to move about so that its respiratory system can be thoroughly irrigated. A shark that has been caught and drugged so it can be transferred to an aquarium must be kept moving by its handlers if it is to recover its normal powers once the effects of the drug wear off. Nevertheless there are certain species living on the ocean floor which swim very little and then not too expertly. In aquariums lesser dogfishes can be seen regularly and rhythmically flapping the margins of their gill openings.

Sharks are endowed with remarkably acute sensory organs which help them to locate prey from afar. The most efficient organs are those connected with the lateral line and those related to scenting. The lateral line extends from the rear of the eye to the base of the tail and consists of a series of long canals situated under the skin, from which small tubes run through a pore to the surface. The canals are filled with mucus and contain numerous nerve cells.

The most recent studies on the sensory perceptions of sharks indicate that waves of high frequency are picked up by the acoustic system and those of low frequency by the lateral line mechanism. But the precise function of the latter has not thus far been completely explained.

When sudden sharp movements break the tranquillity of the water—as when, for example, a fish is struck by a harpoon—the sound waves that are produced are detected by the shark's lateral line apparatus which relays a series of impulses to the brain, alerting the animal without involving its vision. The shark immediately starts describing broad circles or following a zig-zag path towards the point where the disturbance has occurred. It is obvious that the lateral line system is extremely sensitive, conveying vital information to the brain, such as the position, the size and even the swimming speed of the prey, and probably other important facts upon which the success of the hunting foray depends.

When a shark has managed to locate its objective—say a wounded fish—in this manner, it heads straight towards it, partially guided now by the scent of blood. As it swims it swings its head characteristically from side to side, the object being to explore as wide an area as possible in order to pinpoint the exact location of the prey. The position of the shark's nostrils (two grooves placed either longitudinally or diagonally) is of value for they enlarge the surface of contact between the water and the nasal mucus. Examination of the brain of a shark shows that the olfactory lobes are especially well developed.

Many experiments have been carried out in order to discover precisely how a shark detects its prey. Biologists from North America have recorded on magnetic tape the sounds produced in water by a wounded fish and by a man rippling the surface. When such sounds were relayed underwater by means of a magnetophone, sharks immediately hurled themselves at the loudspeaker. Commander Jacques Cousteau carried out similar experiments in the Red Sea to test the scenting powers of sharks. Members of his team first released a coloured substance in the shallow waters of a coral reef to determine the direction of the current and then poured a quantity of meat broth into the water. Watching the results from their underwater hiding place, the divers soon saw

BASKING SHARK
(*Cetorhinus maximus*)

Class: Chondrichthyes
Order: Lamniformes
Family: Cetorhinidae
Length: up to about 40 feet (12 m)
Diet: plankton

Inoffensive species; colour greyish-blue or brownish, paler underneath. Tiny teeth. Well developed gill arches. Mainly found in North Atlantic.

WHALE SHARK
(*Rhincodon typus*)

Class: Chondrichthyes
Order: Lamniformes
Family: Rhincodontidae
Length: up to about 60 feet (18 m)
Diet: plankton

Inoffensive species and largest living fish. White or yellow marks on back and flanks; longitudinal and transverse lines on upper parts. Found in all warm seas.

GREAT WHITE SHARK
(*Carcharodon carcharias*)

Class: Chondrichthyes
Order: Lamniformes
Family: Isuridae
Length: up to 40 feet (12 m) or more
Diet: sea mammals, fishes, turtles, carrion

One of most dangerous of all sharks, potentially a man-eater. Greyish-blue back, whitish belly. Large triangular, serrated teeth. Found in all warm seas, less often in temperate seas, seldom approaching coasts.

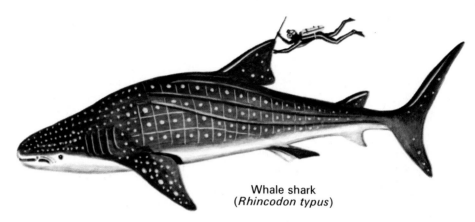

Whale shark
(*Rhincodon typus*)

a number of sharks following the trail, swaying their heads uncertainly from side to side whenever the eddying current momentarily covered the traces of scent. Other tests have convinced the zoologists concerned that sharks are simultaneously sensitive to physical movements such as vibrations and to chemical changes in which both scent and taste play a contributory part.

Now that we know approximately how a shark detects and locates its prey by means of its sensitive lateral line and acute scenting powers, it is easy to understand how accidents involving sharks and underwater divers can occur. In the shimmering half-light beneath the sea a swimmer, having successfully killed a fish with a harpoon gun, will probably tie the trophy to a belt around his waist and swim off in search of another victim. The shark, however, will not only have been alerted by the death throes of the fish but will now be guided unerringly by the smell of its blood. In other circumstances the outline of the diver would probably have frightened it off but now, urged on by hunger and excited by the odour of blood, the shark will be quite fearless, going for the diver's waist and legs. One cannot say for certain that the swimmer would otherwise have been immune, but in this case imprudence and ignorance of shark behaviour will have immeasurably increased the chances of an attack, possibly with fatal consequences.

Inspection of a shark's anatomy shows that the mouth is placed low down on the head, implying that the animal may have difficulty in biting any object not directly below it. Observations indicate, however, that this is not a handicap and that a shark is perfectly capable of sinking its teeth into anything appearing in front of its face or even slightly above it. What happens is that the lower jaw can be extended a little so that the mouth is temporarily in a frontal position. Once the sharp pointed teeth have broken the flesh of a fish, a dolphin or a whale, the predator shakes its whole body vigorously so that the teeth perform a scything action, ripping off a chunk of flesh that may weigh several pounds. Tests in Australia to measure the impact of a shark's bite have shown that an individual 10 feet long can exert a pressure of approximately 50 *tons* per square inch, which is staggering enough to give some impression of the animal's terrible power.

Furnished with such highly perfected sensory systems and armed with such imposing weapons, the killer sharks understandably rank among the most fearsome of ocean predators. Their predominance is in fact contested only by the huge rorquals and humpback whales of the family Balaenopteridae, whose intelligence and capacity for group action are far superior. Even though

Facing page: Lacking a swimbladder (which enables other fishes to float at any desired depth) sharks are compelled to keep constantly on the move. Some species, like the one illustrated in the lower picture, are nevertheless capable of spending much of their time lying motionless on the sandy bottom.

sharks sometimes form packs their attacks are not in any way coordinated, for each animal acts quite independently. Nevertheless, when a shark attacks it does not charge blindly at its victim; before doing so it will usually circle its prey, assessing the situation. As if cognisant of its power and aware that time is on its side, the hunter will swim slowly around, making no sudden movement and not for a moment taking its eyes off its objective. Attracted by the prospects of a meal, other sharks will soon join the fray, describing gradually narrowing circles around the victim until one of them, possibly the hungriest, eventually darts at the wounded or moribund animal, rubbing against it with the rough skin of its back and flanks. This manoeuvre, the last stage before the final assault, is evidently not designed to test whether the prey is sufficiently alert and dangerous to offer resistance, but to find out whether it will make an appetising meal or whether its flesh is too unpleasant. There is no need for the shark to bite the prey or even to bring its mouth into contact with it; for beneath the skin there are sensory pits. The exact nature of these organs is still disputed, but whether they are taste buds or tactile cells they apparently perform the function of distinguishing between what is edible and what is not.

Once this exercise is over and the shark has confirmed that the object in question is edible, it launches its attack, mouth agape, burying its teeth in the soft flesh and shaking its head so as to tear off the largest possible fragment. This leads all its companions to join the attack. For the next few minutes a hunting scene of incomparable ferocity will be enacted as bodies twist furiously,

Since it first appeared in the seas more than three hundred million years ago, the shark has changed little in appearance. Its streamlined body and shape of fins are certain signs of its swimming powers.

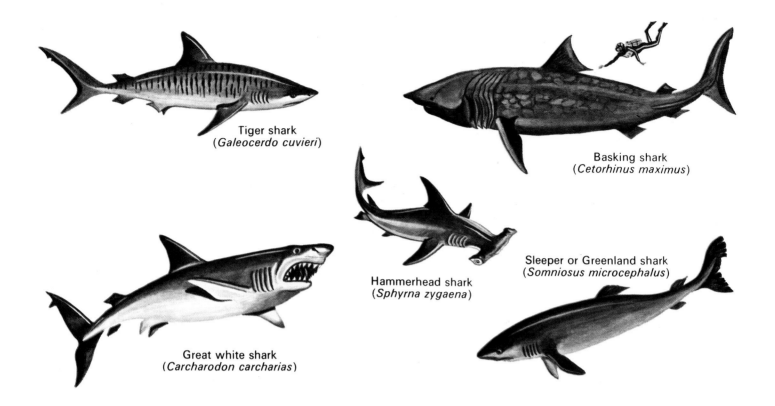

Tiger shark (*Galeocerdo cuvieri*)

Basking shark (*Cetorhinus maximus*)

Hammerhead shark (*Sphyrna zygaena*)

Sleeper or Greenland shark (*Somniosus microcephalus*)

Great white shark (*Carcharodon carcharias*)

tails lash and teeth snap at every last piece of bloodstained flesh. The collective frenzy is of brief duration and when the feast is over the ocean resumes its normal appearance of unruffled calm.

Feasts and fasts

Because there are so many species of sharks, it is impossible to describe a general food pattern. Each group occupies its own ecological niche and feeds in a distinct fashion. Although some species occasionally attack humans, often as a result of their victims' imprudence or miscalculation, not all sharks necessarily go for large prey. Indeed some of them live almost exclusively on plankton, many on fishes, some on molluscs and others on birds, reptiles and sea mammals.

Examination of the stomach contents of sharks provides a long, varied and often surprising list of food substances, which might at first glance seem to confirm the popular belief that they are enormously voracious. But since the animals are poikilotherms (with a variable body temperature) they have no need to expend energy in order to keep their temperature constant and this automatically reduces their food requirements. In Australia aquarium tests have shown that sharks measuring 10–11 feet and weighing about 130 lb consume between 200 and 220 lb of fish in the course of a single year and that they do not eat at all in the cold season.

One astonishing feature of sharks is their ability to store food in the stomach without digesting it. There is a reported case of a shark in captivity which, to the great surprise of those in charge of it, refused its normal food of horsemeat for three weeks. For a while it accepted a few small chunks, only to spit them out shortly afterwards. The shark eventually died and the autopsy showed the presence in its stomach of two perfectly preserved dolphins which it had obviously caught shortly before being captured in its turn. This is all the more strange because the gastric juices of sharks are strongly acidic. Yet there are a number of similar confirmed reports.

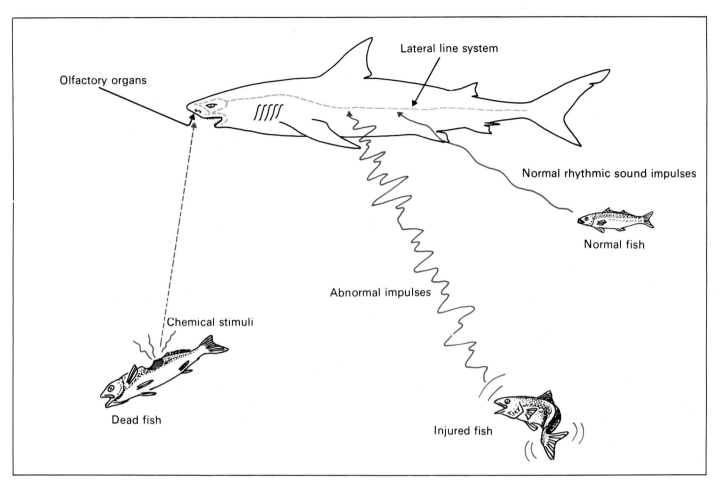

The shark's highly developed sense of smell enables it to locate its potential victims with remarkable accuracy; and detection of prey is evidently reinforced by the so-called lateral line system. Linked with the brain, this consists of a series of long canals filled with mucus and furnished with sensory cells connected by pores to the outside of the body. The system is acutely sensitive to changes in water flow, such as vibrations arising from movements of objects in the water. The nature of the sound impulse will not only permit the shark to locate its prey but also to determine its size and condition – whether, for example, it is healthy, injured or already dead.

To some extent sharks may be considered the scavengers of the seas. They frequently follow in the wake of ships, seizing on any refuse that is tossed overboard. They also track schools of marine mammals, waiting patiently to pounce on a stray elderly or sick individual, a stillborn baby or placental matter. This has been verified each year when grey whales come down from the Arctic to the coastal lagoons of Lower California, where the females regularly give birth. The appearance of the whales heralds the simultaneous arrival in the river estuaries of very large packs of predatory sharks.

The dreaded great white shark

The great white shark (*Carcharodon carcharias*) is, in the opinion of experts, the most dangerous of all species, a genuine man-eater. This enormous animal found in all warm, and sometimes in temperate, seas has an immense mouth with rows of triangular, serrated teeth, each of which may measure up to 3 inches. A formidable swimmer, the shark lives up to the popular reputation for voracity, for the stomachs of several individuals of this species have been found to contain the remains of sea lions weighing close to half a ton and of other sharks more than 6 feet long. Certain great white sharks have even been known to attack boats, for no evident reason. The best known case is that of two Canadian fishermen whose canoe was sunk by a shark during the summer of 1953 off the coast of Nova Scotia. There was no action to provoke the attack of the animal which opened a huge gash in the side of the canoe, tipping its occupants into the water. Although the shark (identified only by a tooth which remained embedded in the boat) made no attempt to attack the men in the water, one of them eventually drowned.

The plankton eaters

The most specialised members of the various shark families are those which, like the whalebone whales, pursue a leisurely path through the oceans in quest of plankton—that seemingly inexhaustible supply of marine food.

As a result of their position in the ocean food chain, very close to that occupied by the primary producers, the plankton-feeding sharks, like the whales, have grown to enormous dimensions. The basking shark (*Cetorhinus maximus*) may, for example, measure up to 40 feet, though 20–30 feet is more usual; and the whale shark (*Rhincodon typus*) which reaches a maximum size of 60 feet is the largest fish in the ocean.

Apart from this specialisation—clearly evident from their unusually small teeth and the presence of gill arches with filaments that serve to filter the microscopic organisms from the water—very little else is known about the behaviour of these two species which, in spite of their gigantic size, are absolutely inoffensive.

According to professional divers, the whale shark is a particularly peaceful species which does not even make menacing motions when faced by daring swimmers who photograph one another gripping the animal's tail or riding astride its back. Yet this tolerant attitude is potentially its most dangerous characteristic as far as man is concerned, for it will sometimes take pleasure in rubbing its back against frail fishing boats. The famous Norwegian explorer and anthropologist, Thor Heyerdahl, in the course of his Pacific journey aboard the raft *Kon-Tiki*, fortunately lived to describe this rather disquieting habit of the whale shark. In his view the animal does it in order to get rid of parasites infesting its skin.

The largest concentrations of basking sharks are to be found in the North Atlantic, yet all that is known is that they come close to shore in summer and cease feeding in winter. Although information concerning this species is extremely sparse, it appears that they reach sexual maturity between the ages of three and five years. This certainly tallies with observations of mating behaviour, following which the animals return to the ocean depths where the females give birth.

Unusual breeding methods

Certainly the most familiar and spectacular aspects of shark behaviour are their hunting activities and techniques. Equally fascinating, yet seldom mentioned, are the ways in which these animals reproduce. In all species fertilisation takes place internally, the male depositing sperm inside the body of the female. The strange fact is, however, that sharks may be either oviparous, ovoviviparous or viviparous, according to species. Among those of the second group, carnivorous tendencies are in evidence as soon as the eggs hatch. The first baby to shake itself free, while still in its mother's genital tract, proceeds to devour one after another of its siblings—a crude yet efficient way of controlling the population.

Many sharks possess sharp triangular teeth and may be dangerous, although liable to attack humans only under certain conditions. Those pictured here are the blackfin shark (*above*) and the great white shark (*below*). The latter is the most dangerous of all sharks.

CHAPTER 14

Wanderers of the ocean

Every year, in the autumn, hordes of freshwater eels (*Anguilla anguilla*) set out on a great journey of adventure. For the past eight to twelve years they will have been living in the lakes, swamps, rivers and streams of Europe, growing a few inches a year. Those that are fully grown and on the brink of sexual maturity are now ready to embark on the long return journey to the waters where they were born. Once arrived it is their destiny to procreate and die.

On very dark nights the females, identifiable by their impressive size (they measure up to 60 inches as against the mere 16 inches or so of the males) allow themselves to be carried by the current to join their companions at the river mouth. Then together they begin their migration westward. Prior to leaving their freshwater habitat they will have fed greedily on aquatic animals, insect larvae, frogs and tadpoles, fishes' eggs and fry, molluscs and crustaceans—stores for the months to come when they will eat absolutely nothing.

Calmly, without any impression of haste, the armies of eels move out to sea, travelling between 15 and 20 miles every day. Those that have departed from rivers flowing into the Mediterranean are joined in due course by others that have travelled from rivers farther to the west. By March the advance guard will have reached their destination in the Sargasso Sea after a non-stop journey which has taken them some 2,500 miles from the shores of Europe. Here, at a depth of about 1,500 feet, where the water has a temperature of 15 °C (59 °F) and a saline content of 36 parts per thousand, the females release up to nine million eggs which are fertilised in the water by the spermatozoa of the males. More eels continue to arrive until the end of June, worn out by the exertions of the journey and the lack of food. Each animal sum-

Facing page: The large fishes which wander the oceans are a rich source of protein for countless millions all over the world. Fishermen in the Mediterranean, shown here landing a tuna, know precisely when shoals are to be found off their shores as well as the traditional migration routes which many species have followed since the beginnings of recorded time.

The common eel and conger eel can be distinguished by the different positions of the mouth and by the outline of the dorsal fin which, in the latter species, begins closer to the head.

mons up the necessary strength to lay its eggs or to fertilise them and then dies.

It takes only a few days for the larvae to hatch. At birth they are tiny transparent fishes, known as leptocephali. Before scientists realised that these were in fact baby eels they were thought to be a distinct marine species.

Soon after they are born, the leptocephali are carried by the Gulf Stream back in the direction of the rivers and streams of North Africa and Europe. It is a journey far longer and more hazardous than that previously undertaken by their parents; and it is to the Danish biologist Johannes Schmidt, who first discovered the eels' spawning grounds in the Sargasso Sea about twenty years ago, that we owe the following details.

At the end of summer the leptocephali, feeding on plankton, are barely 2 inches long and are still close to the American coast. Two months later, drifting on the powerful current flowing from the Gulf of Mexico towards Europe, they are half-way to their destination. Yet it is not for another two years, when they measure about 3 inches, that they finally reach the shores of the Old World.

Now that they are almost at journey's end and close to the mouths of rivers the tiny fishes undergo a metamorphosis. The entire anatomy is transformed; the body, hitherto transparent, takes on a rosy colour, becoming shorter and rounded, and the teeth disappear. At this stage they are known as elvers and they are now ready to begin their freshwater life.

This is when the paths of the young eels diverge. Although many pursue their journey upriver, others remain close to the river mouths and in the estuaries. The interesting point about this is that the sex of the individual eel depends on whether it chooses to settle in the river or by the sea. Of those that head upstream the vast majority are females, while those that remain behind in and around the estuaries are males. It seems, therefore, that the nature of the water in which they live is one of the factors determining their sex. Although the occasional male has been encountered in a freshwater habitat (which would seem to contradict the theory) laboratory examination of such individuals has shown them to have undergone genital mutation (females transformed into males).

On the Iberian peninsula and in North Africa the elvers enter the rivers in October; in France and Ireland it occurs in January, and in England a month later. The eels reach the Baltic and extreme eastern parts of the Mediterranean the following May. Eventually they settle in almost all the rivers of North Africa and Europe with the exception of the Danube basin, for the high hydrosulphuric acid content of the Black Sea is a barrier to further progress.

The journeys upriver take place under cover of darkness and involve enormous numbers. Undeterred by any obstacles in their path, whether they be rocks or waterfalls, the eels battle on. As they advance the size of the invading army is gradually reduced. Small groups disperse up all the tributaries, bringing their epic journey to a conclusion far from their starting point, in a stream, a lake, a pond or a swamp.

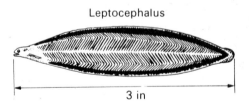

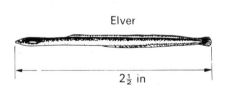

When the larva of the common eel, known as the leptocephalus, reaches the mouth of its home river, it undergoes a complete physical transformation. The body of the elver, as it is now called, becomes narrow and rounded and is also slightly reduced in size.

On wet days some elvers may leave the water and by slithering through the grass find their way into isolated ponds. Usually they remain there until they die, being unable, very often, of making the same journey in reverse.

The female eels remain in the rivers for eight or ten years, feeding voraciously in spring and summer and burying themselves in the mud at the first sign of cold weather. When ready to breed they take on splendid silvery-grey nuptial colours and begin their journey towards the sea. The males reach sexual maturity at four to six years of age.

One extremely interesting problem relating to the reproductive habits of freshwater eels still arouses controversy among zoologists. Both the American and European species spawn in the Sargasso Sea, in zones so very close to one another

The common eel (*above*) and the conger eel (*below*), though superficially alike, differ in habit. The former, despite being born in the sea, is a freshwater species, the latter a marine species.

Geographical distribution of European common eel.

COMMON EEL
(*Anguilla anguilla*)

Class: Osteichthyes
Order: Anguilliformes
Family: Anguillidae
Length: female, up to 60 inches (150 cm)
male, 16–20 inches (40–50 cm)
Diet: in sea, plankton; in river, insect larvae, amphibians, molluscs, fry

The larvae, known as leptocephali, are transparent, flattened and leaf-like. When they reach the river mouth their size decreases, the body becoming rounded and turning pink, at which stage they are called elvers. The eels live in the rivers and lakes for eight to ten years and when sexually mature their back takes on a darker colour with silvery tints.

After living for eight to ten years in rivers female eels reach sexual maturity and begin the long journey back to the traditional spawning grounds in the Sargasso Sea, there to lay their eggs and die.

that the spawning grounds may even overlap. Yet no confusion seems to arise and the two races remain quite separate, implying that each new individual succeeds in finding its way back to the rivers where its parents grew to adulthood. But how can such tiny fishes, originating in different continents yet born together, possibly know whether to head eastwards to Europe or in the opposite direction towards America? According to an apparently logical theory which has been current for some time, the young eels are far from infallible in their sense of direction and pay for such a mistake with their lives. If an American leptocephalus by error heads eastwards it will metamorphose into an elver while still in mid-ocean and since it cannot survive in this new form in such surroundings it will die before reaching Europe. Conversely, the European eel which mistakenly takes the shorter route to America will be incompletely developed when the time comes for it to enter a river and it too will die. So the two populations would appear to retain their individuality by a process of natural selection.

Dr Denys Tucker has put forward another plausible theory. In his opinion it is highly unlikely that both the European and American species reach their spawning grounds in equally good shape for reproductive purposes. Clearly the former are subjected to far more strenuous effort because of the greater distances travelled. Dr Tucker concludes that the European eels never actually reach the Sargasso Sea and that they die in mid-ocean. The leptocephali which subsequently make their way to European shores will therefore have hatched from eggs

The sturgeon, with fins reinforced by cartilaginous rays and its head prolonged by a type of beak furnished with four barbels, feeds largely on insect larvae when juvenile. It spends most of its adult life in the sea, feeding on small fishes, worms and other small invertebrates, returning to rivers only to spawn. Sturgeon's eggs are the raw material for caviar.

laid by the American species. To the obvious objection that two distinct races undeniably exist, this biologist has an answer, namely that if the eels do indeed possess individual characteristics these must be acquired during the earliest period of their life when they are influenced by different natural conditions. But this theory fails to explain why, when they are ready to reproduce, the European eels embark on such a long and (in his view) suicidal journey. One of the more perplexing aspects of this problem is that so far no adult eel in breeding condition has yet been caught at sea.

The sturgeon, provider of caviar

Whereas certain fishes, such as eels, are born in the sea and then migrate towards rivers where they subsequently spend the major part of their adult life, others do precisely the opposite, living in the ocean and only entering rivers to spawn. Among the latter the best known are salmon and sturgeon. These two types, so different from each other, have one feature in common, that both provide man with food. Yet whereas the salmon is appreciated for its appetising flesh, the sturgeon is prized because of its eggs which are used in the preparation of the delicacy known as caviar.

The beluga sturgeon (*Huso huso*), largest of all species, alone provides half of the caviar consumed in the world; but

Geographical distribution of the genus *acipenser*

in fact there are some two dozen species, one of which, the common sturgeon (*Acipenser sturio*), will be described here. Individuals of this species remain in the sea for almost the entire year, normally in shallow waters near the coasts. Towards the end of winter they leave the marine environment and head for the mouths of rivers. The signal for departure is given by the males which are soon followed by the females. They halt for a month or two in the salty waters of the estuaries until their reproductive organs are fully developed; then they swim along the river bed until they reach what is approximately a midway point between source and outlet. Each female selects a suitable site and lays her eggs in enormous numbers, sometimes totalling as many as three million. Simultaneously the males—four, five or even up to ten to each female—release their spermatozoa in the water.

The fertilised eggs stick to the bottom or to water plants, safe from the current. Incubation is short and the eggs hatch some time between the end of March and the middle of June, varying with the locality. The baby sturgeon remain for one or two years in the river, feeding principally on insect larvae, crustaceans and other freshwater animals. At the end of about three months they measure 2–4 inches and at a year old 12–16 inches. The females have a more rapid growth rate than the males. Then they swim down towards the mouth of their natal river, spending another one or two years in the transitional zone of the estuary before venturing out into the open sea.

After the adults have performed their reproductive functions they head back for the sea, completely exhausted, many of them failing to survive the arduous return journey. Those that reach their destination restore their energy by feeding voraciously. Nevertheless when winter comes and storms in the upper courses of the rivers once again create ideal conditions for another migration, these adults are still too weak to undertake the journey and put it off until the following year. Consequently sturgeon only spawn once every two years. Males reach sexual maturity in their eleventh year when they measure a little over 4 feet and are fertile until twenty years old. The reproductive organs of the females are not functional until they are fifteen years old but they are capable of procreating up to the age of thirty years or more.

The world population of sturgeon is currently on the decline, a state of affairs for which man is responsible, mainly as a result of river pollution. This bars the way to the traditional spawning grounds and compels the fishes to lay their eggs in the lower reaches of rivers, where hatching is unpredictable.

The mystery of tunny migrations

From April onwards great shoals of tunny, containing hundreds, sometimes thousands, of individuals, collect off the Atlantic coasts of the Iberian peninsula. Their appearance sparks off feverish activity among local fishermen for this is a regular annual encounter. The massive, streamlined tunny with their narrow, sickle-shaped fins and crescent-like tail are easy to

---

COMMON STURGEON
(*Acipenser sturio*)

Class: Osteichthyes
Order: Acipenseriformes
Family: Acipenseridae
Length: 120 inches (300 cm) and more

Body encased in bony plates arranged in five longitudinal lines. Slender head, ending in a kind of beak. Asymmetrical tail.

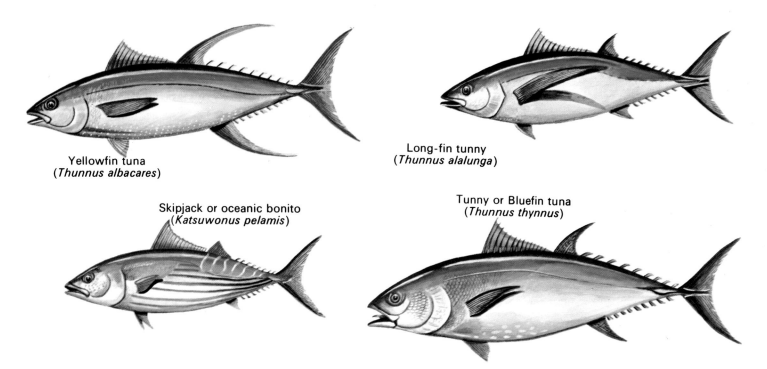

Yellowfin tuna (*Thunnus albacares*)

Long-fin tunny (*Thunnus alalunga*)

Skipjack or oceanic bonito (*Katsuwonus pelamis*)

Tunny or Bluefin tuna (*Thunnus thynnus*)

follow for they swim close to the surface and not too far from shore. Helped by the current, they all head in the same direction—toward the Strait of Gibraltar. Having spent the winter in the Atlantic Ocean age-old instinct now drives them back to their regular spawning grounds. Generation after generation of tunny return predictably every year to their chosen spots in the Mediterranean where the water temperature ranges from 16°C (61°F) to 19°C (66°F). The principal zone of reproduction is in that part of the ocean lying roughly between Sardinia, Sicily and Tunisia; the other is in the waters that separate the south-eastern coast of Spain from the southern shores of the Balearic Islands. To reach the spawning grounds the fishes will have travelled many hundreds of miles. Once arrived, each female lays several millions of eggs which are duly fertilised by the males and left floating in the clear Mediterranean waters. These activities reach a climax in June; then, from July until the end of September, the tunny make their way back by the same route into the Atlantic.

Ever since the dawn of civilisation the fishing people of the Mediterranean have taken advantage of the punctual arrival of the tunny in spring and their disappearance in late summer to reap a rich harvest. Occupied with their livelihoods, it has probably never occurred to them to wonder where the huge fishes spend the remainder of the year. That interesting problem

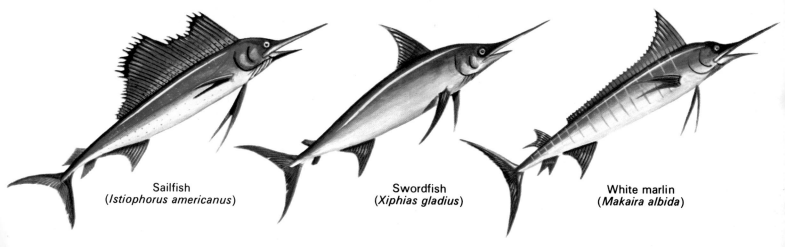

Sailfish (*Istiophorus americanus*)

Swordfish (*Xiphias gladius*)

White marlin (*Makaira albida*)

has been left to the scientists to solve. In recent years exhaustive experiments have thrown some light on the habits of these ocean wanderers, although certain important points of detail remain to be clarified.

In order to find out what happens to the tunny before and after reproduction ichthyologists have resorted to the same methods as have been employed so successfully by ornithologists to trace migration patterns of birds. By catching, marking and releasing individual tunny which may later be recovered by fishing vessels, the scientists have been able to plot the courses of the fishes across the Atlantic and have also gained valuable information about tunny growth and longevity. Thus Portuguese and French fishermen in the Gulf of Biscay have caught tunny which were marked on the American Atlantic coast, indicating that they must have swum at least 3,000 miles. Tunny from Norway have similarly been caught near the Strait of Gibraltar after a nine-month journey of some 1,800 miles.

Although the precise details of the migrations of Mediterranean tunny are still unknown, the general pattern is now clear. When the tunny leave their spawning grounds in the summer and return to the Atlantic, the majority move northward. Their destination is now the North Sea, the waters of which teem with sardines, anchovies, scads, mackerel and many other kinds of small and medium-sized fishes. Here the tunny stay to feed until the end of November. Then, as the water turns appreciably colder, they head south again to complete their annual cycle.

Meanwhile what has happened to the eggs left floating in the sea? Two days after being laid they hatch, freeing the larvae. These survive initially by eating the contents of the yolk sac and later go on to consume zooplankton. In the first year growth rate is phenomenal, for after twelve months the fish measures 24 inches in length and weighs about 9 lb. Then growth slows down considerably. When it reaches sexual maturity, at the age of three years, its size has increased to 40 inches and its weight to 33 lb. Two years later the measurement has gone up to 60 inches and the weight has soared to about 285 lb; and at the age of thirteen years the fully grown tunny measures over 8 feet with a body weight of some 440 lb. Although these figures are impressive they are modest in comparison with those of older individuals that live a solitary existence in the Atlantic. Some of these measure up to 16 feet long and weigh almost 1,700 lb; and in the Bosphorus Strait one tunny was caught which tipped the scales at just under one ton.

Such, in outline, is the life cycle of the tunny or bluefin tuna (*Thunnus thynnus*) which inhabits the Atlantic. Other members of the family undertake equally remarkable migrations, notably the long-fin tunny (*Thunnus alalunga*) of which there are separate races, one in the Atlantic, the other in the Pacific. The former breeds in winter in waters close to the Azores, the Canaries and Madeira, heading northward to Ireland and returning to its spawning grounds in October. The latter spawns around the Midway Islands and then migrates to the Gulf of Alaska. Other individuals of the species travel from east to west across the Pacific from California to Japan.

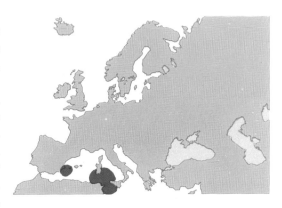

Geographical distribution of tunny in Mediterranean, showing two principal spawning zones.

TUNNY OR BLUEFIN TUNA
(*Thunnus thynnus*)

Class: Osteichthyes
Order: Perciformes
Family: Scombridae
Length: up to 120 inches (300 cm) but some individuals up to 200 inches (500 cm)

Fusiform body; crescent-shaped caudal fin. Back metallic blue; flanks bluish; belly silvery. Migrating species, found in all warm seas.

Facing page: Tunny fishing in the Strait of Gibraltar. Most of the catch goes to the canning industry.

CHAPTER 15

Birds of the open sea

The group of birds known as Procellariiformes might have been specially created for life on the high seas. It is true that birds of other orders, notably frigate birds, gannets and razorbills, lead a life which is partially pelagic, but none of these has adapted to an aerial existence above the oceans so perfectly as the albatrosses, shearwaters, diving petrels and storm petrels–the four families making up the order of tube-nosed Procellariiformes. In fact only during the breeding season do these birds visit dry land and even then they fly long distances over the sea for food.

It is interesting to note that of the hundred or so species of birds comprising this order only the giant petrel (*Macronectes giganteus*), one of the shearwaters, does not rely exclusively on food derived from the sea, for it supplements its marine diet with substances found on land, notably penguins' eggs and chicks, and carrion. All the others feed on animals living at or near the surface particularly those of the macroplankton, with the exception of a few that also eat floating refuse. But basically all these species feed on fishes, marine shrimps, squids and comb jellies. Some albatrosses occasionally eat poisonous jellyfishes as well being among the half-dozen large marine animals to do so.

The various sea birds always hunt prey suitable to their size. Thus the huge albatrosses feed on fishes measuring up to 20 inches and on large squids whose jaws and eyes have been recovered from the birds' stomachs. The storm petrels feed on smaller fishes and the larvae of crustaceans, especially of shrimps; and the shearwaters of the North Atlantic eat herrings as well as smaller species of squid.

Many of the Procellariiformes are nocturnal, which is unusual among sea birds. Doubtless it is an adaptation related to the

Facing page: Albatrosses nest in colonies on oceanic islands or coastal cliffs. Sometimes the nest consists of a simple hole in the ground but more frequently the birds build a mound of mud which is lined with feathers and grass.

Left and facing page: Whereas the courtship ritual of the smaller species of Procellariiformes usually takes place at night and consists of noisy flying displays, that of the wandering albatrosses occurs in daytime and is a mixture of curious dances, mutual greetings, pecks and chirps.

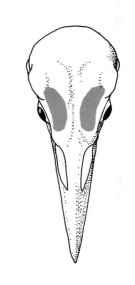

Birds adapted to aquatic life possess so-called salt glands, situated on the head above the eyes. These function as kidneys, helping to eliminate any excess salt consumed with the food.

type of food they eat. In fact plankton and the majority of other marine species undertake daily migrations, rising to the surface at night and remaining there until first light. This is probably the reason why these pelagic birds choose to fish at dusk and during the night, resting by day.

Within the order these birds demonstrate a number of varied hunting techniques. Some albatrosses, having located their prey either by sight or smell, thrust out their neck and harpoon the victim with their hooked bill. Certain petrels, expecially the diving species of the genus *Pelecanoides*, are past masters at catching their prey underwater. Slender and delicate, the birds hover above the waves, extending their long legs in order to balance and brake so that they virtually seem to be walking on the water. Flying against the wind, their movements are slow and deliberate as they spear any tiny animals within range. As for the gadfly petrels of the genus *Pterodroma*, they use surprise tactics, swooping from a fair height and landing alongside their prey, making a catch and then taking off again. The prions (genus *Pachyptila*), classified together with the shearwaters, have an absolutely individual style of fishing, fluttering slowly over the water, trailing their legs on the surface and now and then plunging their head below the waves. When they do so they keep the bill open and since this is furnished with horny lamellae, similar to those of ducks, the birds can filter the water and retain the marine crustaceans and other small organisms which constitute their food.

The Procellariiformes vary considerably in size, ranging from

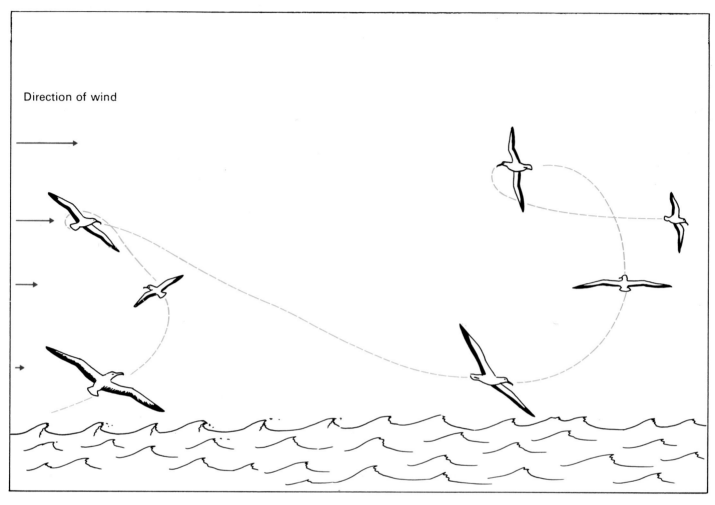

The larger Procellariiformes make use of warm sea currents, differences of pressure and, above all, the force of ocean winds to navigate. They can glide for enormous distances, expending the minimum effort.

the storm petrels which are little larger than swifts to the wandering albatross with its tremendous 11-foot wingspan. The predominant colours of their thick waterproof plumage are grey, white or black. The feet are webbed and the long pointed wings are ideal for gliding, at which they excel, for by making full use of sea winds they can cover great distances with the minimum expenditure of energy. Consequently they exploit all the resources of the high seas, inaccessible to most other birds. The term 'tube-nosed' is derived from the fact that the nostrils of these birds are in the form of short tubes, opening well back or about the centre of the hooked upper bill. Although this is one of the main distinguishing features of the order, the function of this structure has not been discovered. F. W. Jones is one of several ornithologists to claim that the unusually complex mechanism is related to sharpness of scenting power. Tests carried out at sea indicate that petrels scent their way to food and also use their sense of smell to locate their breeding sites and identify their congeners. It may also help Procellariiformes to assess the strength and direction of the wind.

Large or small, the members of this order are distinguished by another feature – their rather unpleasant musky odour. This is due to a yellowish fluid secreted in cells situated in the stomach walls. At low temperature this fluid condenses and takes on the consistency of wax. It appears to serve a variety of purposes. When regurgitated in small quantities through the nostrils in the course of the bird's preening activities, it evidently supplements the oily secretions of the preen glands in waterproofing the feathers. The fluid can also be ejected with some considerable force through the mouth, and in such circumstances provides an effective means of defence, for the rancid odour is enough to deter any intruder. Clearly this action is linked with the regurgitating habit of sea birds in general when disturbed or molested. But the most important and valuable function of this strong-smelling fluid is probably to furnish the young with an additional source of energy-giving food, for the nutritional fat content is some five to ten times greater than that of the fishes from which the substance is derived. Some scientists claim that it is because of this specialised diet that the chicks of these species grow so slowly; for although rich in fats it is deficient in protein. Another theory suggests that the fatty layer forms a protective film around prey which has already been subjected to the action of gastric juices, avoiding the possibility of their being fully digested by the time the parents complete their long journey back to the nest; and it is even possible that the adults, which never need to drink, may benefit from the action of this

The rancid-smelling fluid secreted in the stomach walls of Procellariiformes can be violently ejected through the mouth as a means of defence against predators.

The wandering albatross has the largest wingspan of any bird – eleven feet.

fatty substance which may have a physiological effect similar to that of fresh water.

Splendidly suited to life in the air and in the sea, the Procellariiformes are clumsy on land, many of them simply dragging themselves along on breast and tarsi. This explains why the introduction of predators such as cats, rats and mongooses to islands where these birds breed has often been catastrophic, with some populations almost wiped out.

Young immature birds spend their first year on the high seas, assembling in zones where food is most abundant. Among the smaller species sexual maturity occurs at about three years of age, among larger species between five and ten years. Well before breeding for the first time each pair selects a nesting site which will be used on all future occasions. After a rudimentary courtship display some species—D. L. Serventy has confirmed it for the short-tailed shearwater (*Puffinus tenuirostris*)—make another nest inspection and then fly off, actually mating somewhere out at sea. The species of moderate size breed every year but the wandering albatross breeds every two or three years.

The courtship ritual consists mainly of strident cries emitted by both birds. Each female lays one egg which is remarkably large for her size. It distends the oviduct and cloaca to such a degree that several days afterwards it is evident that the bird

Some typical representatives of the order Procellariiformes. In addition to their tube-shaped nostrils, these birds all have webbed feet but vary considerably in size and in the form of bill, legs, wings and tail.

Magellan diving petrel (*Pelecanoides magellani*)

Leach's petrel (*Oceanodroma leucorhoa*)

Sooty shearwater (*Puffinus griseus*)

Sooty albatross (*Phoebetria fusca*)

Black-browed albatross (*Diomedea melanophris*)

Common diving petrel (*Pelecanoides urinatrix*)

Fulmar (*Fulmarus glacialis*)

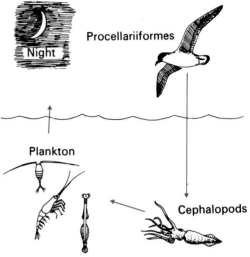

Plankton, together with the cephalopods that feed on it, rise to the surface of the sea during the night. The tube-nosed Procellariiformes accordingly adapt their feeding habits to this rhythm and hunt in hours of darkness.

The giant petrel nests on land and its breeding area extends to the desolate south polar regions.

has recently laid. Sexual dimorphism is hardly apparent, but males are generally a little larger than females.

The incubation period is comparatively long, ranging from forty days in the smaller species to eighty days among the larger ones. The duration of rearing the chicks also varies from about two months in the former groups to ten months, for example, in the case of the wandering albatross.

The Procellariiformes, unlike the majority of land birds which are capable of advancing or delaying their nesting dates, according to weather conditions, invariably breed at a precise time of year. A. J. Marshall and D. L. Serventy, in their studies of huge colonies (often millions strong) of slender-billed shearwaters in Tasmania and southern Australia, have pointed out that the females lay their eggs, with astonishing synchronisation, between the 24th and 26th November. This behaviour is more readily achieved in birds nesting in close colonies because the pairs stimulate each other, producing a breeding crisis.

Before completing their growth the chicks accumulate a huge

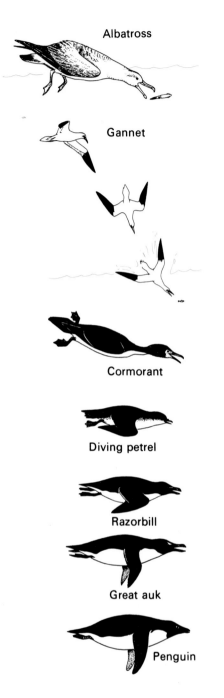

Sea birds, to whichever orders they belong, have adapted in various ways to aquatic life and exemplify different fishing techniques. Between the styles of the huge gliding albatrosses and the flightless penguins, there is a wide range of intermediate techniques.

The storm petrels of the southern hemisphere have a highly individual way of collecting the plankton on which they feed, flying slowly above the surface so that they appear to be walking on the water.

quantity of fat so that they are even heavier than the adults (80 per cent heavier in the case of the slender-billed shearwater, about 50 per cent in other species). The advantage is that the fat apparently helps the chicks to survive during the long absences of their parents, away looking for food, protects them against inclement weather and enables individuals that may be abandoned by the parents to grow to maturity. This occurs quite frequently for both adult males and females often leave the breeding colonies before the baby birds are fledged. In such cases the young birds have to complete their development, abandon the nest and begin their fishing activities without any guidance from the adults, testifying to the important role of innate, instinctive behaviour among these species.

The Procellariiformes make up one of the most primitive groups in the bird kingdom. There is evidence to indicate that they originated in the southern hemisphere and their closest relatives are certainly the penguins, which have a similar geographical distribution. Both the tube-nosed sea birds and the penguins are better adapted to aquatic life than any others.

Ornithologists divide the order Procellariiformes into four families. The family Diomedeidae is represented by the albatrosses, including the huge wandering albatross. The largest family, Procellariidae, groups the shearwaters, some petrels (including the giant petrel) and the prions or whalebirds. The storm petrels belong to the family Hydrobatidae and the diving petrels to the Pelecanoididae.

This drawing shows the different ways in which various sea birds procure their food, either by depriving other birds of their prey in mid-air, by swooping down and seizing plankton or fish at or just below the surface, or by diving into the sea and catching prey at greater depths. The species depicted here are: 1. Great skua. 2. Pomarine skua. 3. Giant petrel. 4. Pintado petrel or Cape pigeon. 5. Antarctic prion. 6. Storm petrel. 7. Skimmer. 8. Herring gull. 9. Black noddy. 10. Frigate bird. 11. Gannet. 12. Tropic bird. 13. Brown pelican. 14. Tern. 15. Greater shearwater. 16. Albatross. 17. Phalarope. 18. Guillemot. 19. Penguin. 20. Diving petrel. 21. Common cormorant. 22. Guanay cormorant. 23. Scoter.

CHAPTER 16

The mighty whales

An ancient book of travel tells the story of a group of sailors drifting helplessly in a small boat who finally landed on an empty island. They decided to break up their boat as firewood, but no sooner had they managed to light a fire when their 'island' suddenly started to move. Within seconds the petrified men had been dragged down to their doom by the huge whale whose peace they had unwittingly disturbed.

Whales that have been mistaken for islands feature in the tales of many seafaring nations, and understandably so, for they are gigantic animals, the largest that have ever appeared on earth. The weight of a blue whale is equivalent to that of about 1,600 men, to 150 large cattle or to 25 elephants; and the sperm whales, although smaller, are still more massive than any land mammal. Such enormously heavy creatures would be quite incapable of moving about on land and indeed a whale stranded on a seashore is doomed, dying of asphyxia from the weight of its own body compressing the lungs. But the laws of proportional size and weight (which dictate that no animal much larger than an elephant can exist on land) are not applicable in the ocean, so that mammals as immense as whales can move about in this element with astonishing freedom and agility.

All the representatives of the order Cetacea, which includes whales, dolphins and porpoises, are marine mammals, or more precisely aquatic mammals, given that some dolphins live in rivers. Large or small, with their short neck, flippers and muscular tail, the cetaceans are easily mistaken for fishes. In fact the eminent Aristotle, together with almost all naturalists up to the eighteenth century, never doubted for a moment that the animals were true fishes. Nevertheless they are warm-blooded mammals, bearing live young.

Facing page: The beluga or white whale is one of the toothed whales, measuring 12–13 feet long. Living chiefly in northern circumpolar regions, it feeds on fishes, crustaceans and molluscs.

Skeletal structure of the forelimbs of a primitive land mammal, a sea-lion and a dolphin, showing the progressive shortening of the central bones and the elongation of the fingers which eventually support the flippers.

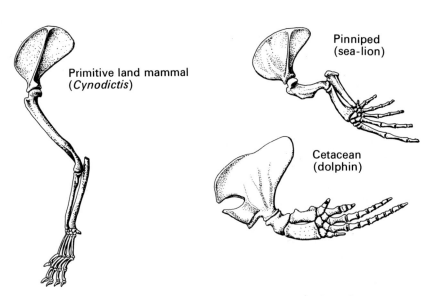

Toothed and toothless whales

Because they spend their entire life in water, cetaceans are mammals at the farthest remove from the primitive insectivore stock. The enormous head seems to be attached without a neck to the rest of the smooth, naked body, the line of which is broken only by a pair of flippers. These are transformed forelimbs and still retain the skeletal structure of the hand. Some species also have a small but boneless dorsal fin. The large tail ends in a fin that is flattened horizontally instead of vertically, as in fishes. The hind limbs have vanished but there are vestiges of the pelvic girdle in the Greenland right whale. Hairs are confined to a few scattered sensory vibrissae in the head.

The teeth have also been modified for since food cannot be chewed under water the specialised dentition of land mammals is of no value. Among the toothed cetaceans (Odontoceti) the individual teeth are undifferentiated, used only for gripping food. The baleen or whalebone whales (Mysticeti) have no teeth at all. The two groups differ in other ways too so that some zoologists are inclined to think that they may have stemmed from separate species, any similarities being due to convergent evolution. If this is accepted they clearly belong to different orders, but for present purposes the orthodox classification will be retained.

Cetaceans have to solve the same problems as are encountered by the pinnipeds, particularly those relating to respiration, temperature regulation and resistance to high pressure.

The embryo of a whale, measuring less than half an inch, possesses nostrils (like all mammals) at the tip of the snout. Later, by the time the fetus is an inch long, they have moved to the normal position high on the top of the head. These nostrils (of which the Mysticeti have two and the Odontoceti only one) are known as blow-holes. When the whale surfaces it releases the air contained in its lungs which then forms a cloud of particles of mucus—not, as is popularly believed, a jet of water—which falls back in the form of droplets. An expert is able to identify the species of a whale simply by watching the distinctive shape and force of this cloud.

Whales breathe atmospheric air using their lungs (whereas fishes use gills to absorb oxygen dissolved in water). Since the animals are capable of remaining under the surface for twenty or thirty minutes at a time it might be assumed that they possess enormous lung capacity but in fact this is not the case. The lungs of whales are of normal size in relation to the animals' bulk, and those of sperm whales, rorquals and some others are proportionately smaller than those of land mammals. Nevertheless the cetaceans do have another advantage when it comes to respiration. It has been calculated that every time a whale breathes in it replenishes 80 to 90 per cent of the air in its lungs, whereas in man, for example, there is only 10 to 15 per cent of fresh air introduced. In addition a whale's reserves of oxygen are more abundant and distributed in a far more efficient manner; thus almost half (41 per cent) of the gas is stored in the muscles, another 41 per cent in the blood, 9 per cent in the tissues and only 9 per cent in the lungs. In man, by contrast, a mere 13 per cent of the total oxygen is stored in the muscles and 34 per cent in the lungs.

Since whales (particularly sperm whales) sometimes dive to considerable depths—3,000 feet and more—accidents arising from high water pressure occasionally occur. There have been at least two reports of blue whales having suffered a fracture of the tympanic bulla (situated in the skull) although the injury had healed by the time the whales were caught. Nevertheless a delicate network of blood vessels (comparable to the rete mirabile in the neck of a giraffe but here much more complex and distributed over the entire body) guarantees a regular supply of blood to the brain and also keeps the blood pressure within tolerable limits.

The third important problem is that of temperature regulation. Whales are warm-blooded animals and have a heart with four chambers to keep the blood circulating. But since they live in cold water it is vital that they do not lose too much heat, and the inside of the body must be protected. They lack body hair and the skin is thin and delicate, except in some species such as sperm whales and belugas where it is leathery. A thick layer of fat, known as blubber, situated under the skin, therefore helps to insulate the body and compensates for the absence of external protection. In seasons when food is most abundant the blubber may be 24 inches thick in the right whales, 8 inches in the sperm whale and a little less in other species, but the thickness varies in different parts of the body.

Cetaceans, large and small

According to the 1967 classification of Dale W. Rice whales may be divided into two major groups, Mysticeti and Odontoceti. The former order is made up of three families, the latter of seven. The families of toothless or whalebone whales are Balaenidae (right whales), Eschrichtiidae (grey whales) and Balaenopteridae (rorquals and humpbacks).

There are three species of right whales, so named simply because at one time they were the right whales to hunt; in all

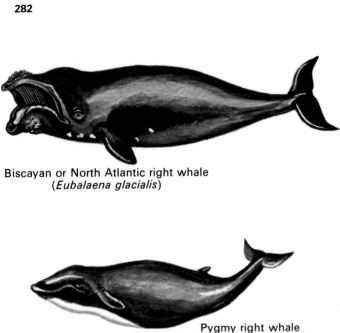

Biscayan or North Atlantic right whale
(*Eubalaena glacialis*)

Bowhead or Greenland right whale
(*Balaena mysticetus*)

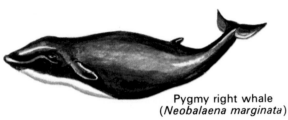

Pygmy right whale
(*Neobalaena marginata*)

Common rorqual or fin whale
(*Balaenoptera physalus*)

Rudolph's rorqual or sei whale
(*Balaenoptera borealis*)

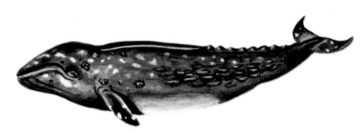

Californian or Pacific grey whale
(*Eschrichtius gibbosus*)

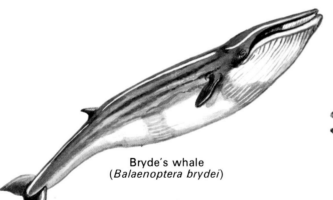

Bryde's whale
(*Balaenoptera brydei*)

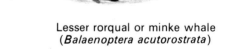

Lesser rorqual or minke whale
(*Balaenoptera acutorostrata*)

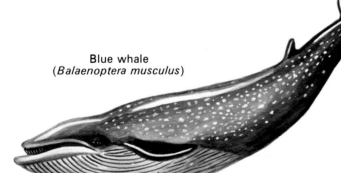

Blue whale
(*Balaenoptera musculus*)

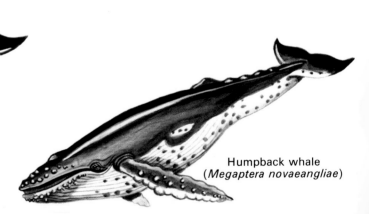

Humpback whale
(*Megaptera novaeangliae*)

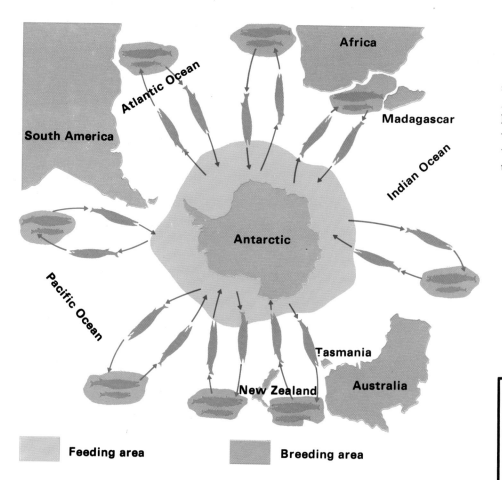

Feeding area **Breeding area**

Whalebone whales follow an annual rhythm of migration. They spend the summer, both in the northern and southern hemisphere, in the Arctic and Antarctic respectively, feeding mainly on krill. In winter they return to warm and temperate seas to reproduce, after which they head back for the higher latitudes to fatten again.

of them the throat is completely smooth, without any grooves. The Greenland right whale or bowhead (*Balaena mysticetus*) may reach a length of 60 feet. Its body is protected by an unusually thick layer of blubber, as is the case too with the Biscayan or North Atlantic right whale (*Eubalaena glacialis*). The latter is in fact found in most oceans and has been the object of some scientific controversy, for although certain experts distinguish it from the very similar black right whale (*Eubalaena japonica*), of the North Pacific, and southern right whale (*Eubalaena australis*), others regard the three as belonging to the same species. The smallest member of the family is the pygmy right whale (*Neobalaena marginata*) which does not grow larger than 20 feet and is found only in the seas of the southern hemisphere where it has apparently never been abundant.

The Californian or Pacific grey whale (*Eschrichtius gibbosus*) is the only member of its family. It measures up to 45 feet and there are two or four grooves on the throat. Although it is now found only in the North Pacific, bones of the species have been recovered from the Dutch coast, showing that it may also be, or have been, an inhabitant of the Atlantic.

The rorquals differ slightly from true baleen or whalebone whales in that their throat is completely grooved, this structure enabling the mouth to be stretched to an unusual extent. The largest of the rorquals is the blue whale (*Balaenoptera musculus*), which used to reach 100 feet or more; but both the common rorqual or fin whale (*Balaenoptera physalus*) and Rudolph's rorqual or sei whale (*Balaenoptera borealis*) are also huge;

RORQUALS

Class: Mammalia
Order: Cetacea
Family: Balaenopteridae
Diet: basically small crustaceans (krill)
Gestation: 10–12 months
Number of young: one

BLUE WHALE
(*Balaenoptera musculus*)

Total length: average 80 feet (24–25 m), maximum 104 feet (31 m)
Weight: up to 130 tons

Largest and heaviest of all mammals that have ever lived, either on land or in sea. Numerous longitudinal grooves on throat. Back and belly uniformly bluish-grey; lighter marks irregularly distributed over body. Once present in all oceans, especially abundant in Antarctic. Protected species.

FIN WHALE
(*Balaenoptera physalus*)

Total length: average 65 feet (20 m), maximum 80 feet (24–25 m)
Weight: up to 75 tons

Second largest rorqual species, with proportionately smaller flippers but larger dorsal fin than blue whale. Back greyish, belly white. Species hunted by authorisation, but in great danger of extinction.

SEI WHALE
(*Balaenoptera borealis*)

Total length: average 50 feet (15 m), maximum 60 feet (18 m)
Weight: up to 60 tons

Back bluish, belly white; small flippers; particularly soft baleen plates. Species much hunted and endangered.

The whalebone or toothless whales are magnificently adapted to ocean life. Their reproductive processes are similar to those of land mammals and the mother suckles her baby on milk for several months after its birth. The calf grows very rapidly.

the former averages 65 feet with a maximum of 80 feet, the latter 50 feet and 60 feet respectively. Bryde's whale (*Balaenoptera brydei*) is somewhat smaller, on average 40 feet long. The smallest representative of the Balaenopteridae, however, is the lesser rorqual or minke whale (*Balaenoptera acutorostrata*) which is a modest 30 feet in length.

Separate mention should be made of the humpback whale (*Megaptera novaeangliae*), averaging 40 feet and growing to a maximum of 50 feet. It differs somewhat from other Mysticeti, its distinctive feature being the size of the flippers. It is a very playful species and may often be seen leaping from the water, huge fins outspread. The Spanish zoologist Cabrera writes that a humpback is as large as three elephants.

The seven families of toothed whales, some already described in a previous chapter, are Delphinidae (dolphins and whale-named dolphins), Ziphiidae (bottle-nosed or beaked whales), Physeteridae (sperm whales), Platanistidae (river dolphins), Phocaenidae (porpoises), Monodontidae (narwhal and beluga) and Stenidae (long-snouted dolphins).

Whales belonging to the Delphinidae include the grampus or killer whale (*Orcinus orca*), the false killer whale (*Pseudorca crassidens*), the pygmy killer whale (*Feresa attenuata*) and the pilot or caa'ing whale (*Globicephala melaena*), this being the largest, measuring up to 28 feet. All have a dorsal fin and a prominent, swollen forehead.

The Ziphiidae possess an elongated rostrum or beak, longitudinal grooves in the throat region and few functional teeth (except in one species there are only one or two pairs in each

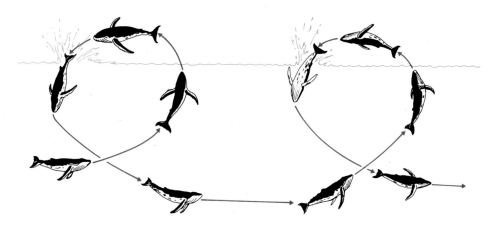

The humpback whale, like other cetaceans, has strongly playful instincts. A favourite exercise appears to be leaping and diving in large circles.

jaw. These whales feed mainly on cephalopods. The bottle-nosed whale (*Hyperoodon ampullatus*), the male of which measures up to 30 feet, is found in the North Atlantic. Baird's beaked whale (*Berardius bairdii*), an inhabitant of the North Pacific, is the largest member of the family; and less well known are the species of the genus *Mesoplodon*, including Gervais' beaked whale (*Mesoplodon europaeus*). Oliver's beaked whale (*Tasmacetus shepherdi*), from New Zealand, is the species with a larger number of teeth in either jaw.

The two representatives of the Physeteridae are the cachalot or sperm whale (*Physeter catodon*) and the pygmy sperm whale (*Kogia breviceps*), both with functional teeth only in the lower jaw. The Monodontidae consist of two species, the narwhal (*Monodon monoceros*), with its single jutting tusk, and the beluga or white whale (*Delphinapterus leucas*).

The giant blue whale

Linnaeus would appear to have had a sense of humour in applying the Latin word 'musculus', literally 'little mouse', to the blue whale. For this gigantic mammal can measure upwards of 100 feet and weigh up to 130 tons, and is not only the largest living animal but the largest that has ever existed.

Like all the whalebone species the blue whale undertakes regular migrations, motivated by two urgent impulses—the search for food and the mating instinct. Every year, in spring, these whales journey towards the poles. At this season they feed in vast quantities and accumulate a large amount of body fat. By autumn they are in peak condition and turn back in the direction of the tropics to reproduce. In the course of their journey they eat very little and lose a considerable quantity of blubber, having no need for this protective fatty layer in warmer seas.

Surprising as it may seem, these giants of the world's oceans feed mainly on tiny animal organisms not more than an inch or two long, especially the small prawns known as krill (*Euphausia superba*) which form immense shoals in the seas of Antarctica. But how do they catch such very small prey? The stomach of a blue whale is capable of holding up to one ton of krill and obviously the huge animal does not take them one by one for such a feeding rhythm would be far too slow to satisfy its great appetite. Nor can it devour thousands of shrimps at a single gulp by simply opening its massive mouth, for this would entail swallowing large amounts of water at the same time. These feeding problems are resolved by the presence in the mouth

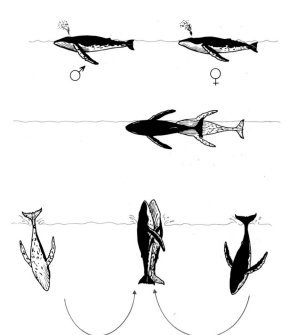

Various phases of the courtship ritual among humpback whales.

of blades or plates which are designed to expel unwanted water and to retain only the food substances. These horny plates of baleen or whalebone (once widely used for corsetry and for other industrial purposes prior to the use of steel and plastic) are arranged in two closely packed rows on either side of the upper jaw. Triangular in shape and separated from one another by only small spaces, the plates hang down from the roof of the mouth and the inner edges become frayed, so forming a filter. When the whale is swimming below the surface in the middle of a shoal of krill it opens its mouth and engulfs a mixture of shrimps and sea water. The tongue then presses against the palate and forces the water out through the filtering baleen plates, isolating the food. Then by some mechanism not yet understood, but plainly making use of the tongue, the plankton is driven through the narrow gullet into the esophagus, there to be conveyed to the stomach.

The Soviet zoologist B. A. Zenkovich has studied the feeding habits of the blue whale in detail and has indicated that it consumes a larger quantity of phytoplankton (of plant origin) than has commonly been assumed. Furthermore he has demonstrated that the species feeds most intensively in colder waters and that larger individuals and gravid females possess a bigger appetite than young whales and non-reproductive females. According to this author a blue whale may consume up to four tons of plankton a day in the Antarctic; and since it remains in this region for some 120 days, simple mathematics will show that a whale is capable of eating more than a million pounds of tiny organisms in a single season.

Not all the whalebone whales have an identical form of diet, and to some extent this is determined by the arrangement and size of the baleen plates. Bryde's whale is especially interesting in this respect; equipped with small, rigid plates, this species feeds mainly on fishes. Examination of the stomach contents of other species, however, has revealed the frequent presence of fishes, cephalopods, worms and copepods—small animals consumed in immense quantities by the Minke whale, which is furnished with very narrow baleen plates.

In March or April, the Antarctic autumn, ice begins to hem in the fishing grounds of the blue whale, the humpback whale and other rorquals, and they now set out on their journeys to warmer climes. Many such whales, on leaving the Antarctic, are encrusted with masses of tiny diatoms which give their body, particularly the belly region, a yellowish hue. These crusts are shed in the course of the northward journey. When the whales begin their voyage they are fairly fat but when they return the following season they are much thinner, implying that they eat relatively little during their sojourn in tropical and subtropical seas. Supporting this theory is the fact that a number of whales tend to hover throughout the year on the edge of the pack ice and that these animals evidently lose little weight. Some authors claim, however, that there are certain areas in warm oceans where the whales gather to restore their energy, notably in the Gulf of Aden, off the coast of Dakar, in West Africa, and in the Caribbean.

RIGHT WHALES

Class: Mammalia
Order: Cetacea
Family: Balaenidae

GREENLAND RIGHT WHALE
(*Balaena mysticetus*)

Total length: maximum 60 feet (18 m)

Skin of adult black, paler in young. Distinctive white mark below chin and front part of lower jaw. Skin of throat smooth as in all right whales. Very long, flexible baleen plates. Not pronouncedly migratory, seldom leaving Arctic regions. No longer hunted.

NORTH ATLANTIC RIGHT WHALE
(*Eubalaena glacialis*)

Total length: maximum 60 feet (18 m)

Completely black, but sometimes skin parasites give belly and other parts of body whitish hue. Head smaller than that of Greenland species, with baleen plates less developed. Wide distribution. No longer hunted.

PYGMY RIGHT WHALE
(*Neobalaena marginata*)

Total length: maximum 20 feet (6 m)

Skin black with scattered grey patches. Small dorsal fin, not found in other right whales. Inhabitant of waters of southern hemisphere. No longer hunted.

The Pacific grey whale, only member of its family, is one of the whalebone whales. It is a coastal species, living in the North Pacific, which makes seasonal migrations up and down the shores of Asia and the west coasts of North America. For some years it has been a protected species.

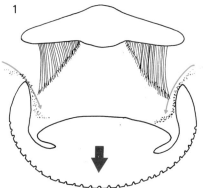

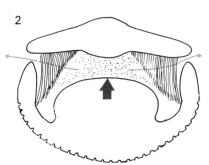

The whalebone whales feed in a specialised fashion. When the animal finds itself in a shoal of krill it opens its mouth (1) and takes in a mixture of food and water. The tongue moves up to the palate, forcing the water out through the filtrating baleen plates (2), retaining the food.

The two-ton baby

Very little is known about the sexual behaviour of the toothless whales but the few observations that have been made suggest that the animals are monogamous and that a courtship and seduction ritual occurs prior to mating. Among humpback whales the male pursues the female, puffing at her until she lies sideways in the water and then extending himself beside her with his belly turned towards her. Both animals then perform a nose-dive, leaving only the tips of their tails showing above water. After that they rear up to the surface once more so that the front part of the body juts out, ending up by coupling in this position. Rorquals evidently proceed in a similar fashion.

Blue whales apparently mate either in May or June in warm oceans and it is in these regions that the females subsequently give birth, approximately a year afterwards (gestation lasting about eleven months). In the majority of species there is only one calf, twin births being rare.

George L. Small, who has made a comprehensive study of the biology and behaviour of the blue whale, points out that nobody has so far seen, let alone been able to measure, the size of a

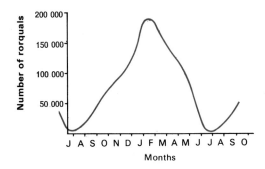

Rorquals spend the coldest months of the year near the tropics and with the approach of summer migrate to the polar regions in order to feed copiously. This graph (prepared by Mackintosh and Brown) shows the number of individuals counted in the Antarctic during the months of a single year. It is evident that a number of whales, presumably immature ones, remain in the polar regions throughout the year.

Facing page: The giant whalebone and sperm whales have been hunted for centuries, but the real impact on their numbers only began to be felt from the eighteenth century onwards when nations realised their true economic value. Indiscriminate hunting led to the near-extinction of certain species and it is only in the last few decades that the massacre has been checked by international agreement. Despite prohibitions relating to certain species, including blue, grey, right and humpback whales, others are still threatened, notably the fin and sei whales.

newly born calf and that conditions make it unlikely that anyone will be able to do so in the future. But examination of the fetus shortly before birth indicates that the calf enters the world already pretty well developed as a miniature replica of the adult, measuring at least 23 feet and weighing approximately two tons.

The calf is suckled by the mother for almost seven months, her milk containing a higher proportion of fats and proteins than that of land mammals, but less water. As the calf feeds under water, the problem of suckling might seem fraught with difficulties, especially since the mouth is adapted for filtration. But the milk is ejected with great force, in the form of an intermittent jet, by the lacteal ducts through the nipples (situated in folds of skin on the side of the vaginal canal), so that it finds its way directly into the baby's throat, by-passing the mouth cavity.

The calf grows rapidly and when it is weaned measures about 50 feet and weighs approximately 23 tons, indicating a daily growth rate of roughly two inches and a simultaneous weight increase of 200 lb. It has not been possible to evaluate the milk secretion of the nursing female but it certainly exceeds 300 lb per day.

While she is suckling, the adult female cannot again be fertilised. Like other Balaenopteridae, therefore, she only gives birth once every two years. Furthermore, if one takes into account the fact that she attains sexual maturity between the ages of four and six years, and that she will not live for more than thirty or forty years, she is unlikely to bear more than about ten calves in her lifetime. Such a comparatively low birth rate would be sufficient to maintain the numbers of the species (which have no natural enemies save the killer whale) were it not for the tragic intervention of man.

The impact of whaling

The tragedy of the whalebone whales is that very early in his development man discovered that these massive animals could be exploited in a variety of ways. In fact every part of their body can be profitably used. Whale oil, extracted from the thick layer of insulating blubber, was formerly used as fuel for lamps and is now an important ingredient in the manufacture of margarine, soap and synthetic resins. Whale meat is eaten in Japan and in parts of western Europe; alternatively it is converted into fertiliser or into food for pets and livestock. The bones are used for making gelatine and glues. The substance misleadingly termed whalebone was at one time widely used in corsetry but nowadays to a lesser extent. The connecting fibres of the tissues make gut for tennis rackets; and even the internal organs of the whalebone whales have been used for the extraction of hormones and vitamins.

Man's encounter with whales evidently began at the dawn of human history, as is proved by the discovery of cave drawings of the animals in Scandinavia. The oldest accounts of whaling in Europe date from the ninth century but the real pioneers

were French and Spanish Basques, and figures of whales, of whaling vessels or of hunters appear on a number of Basque family coats-of-arms. Until the nineteenth century whaling was a dangerous and much romanticised activity. The sailing ships used for whaling were heavy and slow-moving. When the familiar cry of 'There she blows!' went up from the lookout in the crow's nest, the crew launched their frail rowing boat and set off in the direction of the quarry. The harpooner, usually the strongest member of the crew, would stand in the bow, poising his harpoon – a kind of spear at the end of a long rope. Not until he was almost on top of the whale did he release the weapon, thrusting it as deeply as possible into the animal's side. Once the harpoon was embedded in the whale the rope was let out and the chase would begin. The wounded animal was pursued

for as long as necessary until it was sufficiently exhausted for the boat to close with it and allow the harpooner to kill it with repeated thrusts of his lance.

The carcase of the whale was then lashed to the side of the ship. The whalebone was cut out, the blubber stripped (flensed) and the remains of the dead animal left, as often as not, for the sharks and gulls. Sometimes, however, the whole animal was towed to shore to be dismembered on dry land.

Prior to the nineteenth century the modest speed of the whaling vessels and the inaccuracy of hand harpoons set a limit to man's destructive powers. The contest was about even but it was the Biscayan or North Atlantic right whale whose numbers dwindled significantly at first, although later the humpback whale and the sperm whale were hunted. The Biscayan as well as the Greenland right whale were the most frequent victims for they were slow in flight and tended to float after being killed. The rorquals, much swifter, and whose carcases sank immediately, were less often hunted.

The appearance of steam vessels and the invention, in 1864, by the Norwegian Svend Foyn, of a cannon-fired grenade harpoon, radically altered the situation. This weapon has been used, with little modification, until the present day. The cannon shoots a huge harpoon measuring 6 feet and weighing 150 lb. The head is filled with explosive. Three seconds after the harpoon hits home a time fuse ignites the explosive, often killing the whale instantaneously. Then air is pumped through a rubber tube, inflating the carcase of the whale so that it floats.

At the beginning of the twentieth century a way was found to hoist the whale directly onto the ship by means of a slipway in the stern so that the animal could be flensed and dismembered on the spot. This marked the introduction of factory ships, enabling the crews to spend long periods at sea and to kill more whales in a single voyage. Nowadays location of the whales has been perfected by means of helicopters and sophisticated sonar devices.

The widespread use of these new hunting techniques soon led to the decimation of whale populations. It was not long before there were so few whales remaining in Arctic waters that it was not economical to send vessels such a distance to catch them. In fact, Arctic whaling has never been important in the present century. Consequently whalers from Norway, Germany, Holland, Britain, Japan, Russia, America, and other interested nations switched their attentions to the immense whale populations of southern seas. Repercussions were swift and savage. Sperm and whalebone whales were hunted mercilessly in vast number. None was spared, the massacre extending indiscriminately to gravid and nursing females as well as young. Numbers decreased alarmingly.

In 1932, as a result of overfishing, the first international attempts were made to limit whaling activities. In 1946 the International Whaling Commission was established, to which eighteen nations are today affiliated. The Commission prohibited all whaling in certain demarcated zones. It also forbade the killing of whales under a given size and of females accompanied

Facing page: The regulations accepted by many nations allow Japan to hunt certain species of whalebone whales since she is a major consumer of whalemeat.

by their calves. Finally it set a limit to the total number of whales to be caught in a single season, appointing a team of inspectors to make sure that the regulations were obeyed.

One species threatened with possible extinction was the blue whale. At the beginning of the present century the worldwide population count was in the region of 150,000; by 1963 it had fallen to 2,000 or less. Thanks to international protection the numbers had climbed back to 13,000 by 1973. Salvation of the species was only made possible by enforcing the most stringent prohibitions in certain areas. In the North Atlantic the blue whale has been fully protected since 1960 and in the Antarctic since 1965. Two years later the restrictions were extended to cover all the world's oceans. In 1972 the Commission renewed the regulations indefinitely to cover blue, grey, right and humpback whales all over the world.

N. A. Mackintosh has estimated that if the killing of blue whales had ceased as far back as 1945 they would by now have made good their losses and restored their numbers to the turn of the century level. Even if no more are killed in the meantime it is unlikely, indeed impossible, for the blue whale population to get back to the 50,000 mark by the year 2000. According to the same author, the next most threatened species is the common rorqual or fin whale, for this is still hunted. For the 1973-74 season the killing of 1,450 fin whales was authorised on condition that within the next three years there should be a total prohibition on hunting the species below latitude 40°S. There were once some 250,000 fin whales in the southern seas, yet by 1965 these had been reduced to about one-fiftieth of their former strength. If killing is halted they should restore their numbers rapidly. As for other species, although their situation causes much concern, it does not yet appear to be desperate.

Enough food in the Antarctic

The tragic disappearance, within the space of a few years, of thousands of whales which consume countless tons of plankton has inevitably had its effect on the availability of this indispensable source of food. Mackintosh has calculated approximately how much krill which would ordinarily have been eaten by whales now remains available for other predators, placing the annual figure at not less than 30 million tons and probably much more. Assuming protective policies are continued, this figure of surplus food is likely to be halved by the year 2000, even if other krill consumers increase in number.

According to Mackintosh the minke whale, the crab-eater seal, sea birds, fishes and cephalopods benefit most from the dwindling numbers of large whales. Yet no increase has been recorded in the population figures of the first two species and even the various species of sea birds have not increased their numbers to the extent that might have been expected. Only the populations of pelagic fishes and squids seem to show consistent growth; but short of a catastrophe, even if the large whales do return in their former numbers there will always be enough food and to spare in Antarctic waters.

SPERM WHALE
(*Physeter catodon*)

Class: Mammalia
Order: Cetacea
Family: Physeteridae
Total length: males maximum 60 feet (18m)
Weight: 50 tons and more
Diet: cephalopods, fishes
Gestation: probably about 16 months

Enormous head, occupying one-third of length of body, quadrangular in profile. Lower jaw contains two rows of 20–30 pointed peg-like teeth. Small flippers; no dorsal fin but low, irregular crest towards rear of body. Caudal fin may measure 15 feet across. Colour bluish-grey, lighter on belly.

PYGMY SPERM WHALE
(*Kogia breviceps*)

Class: Mammalia
Order: Cetacea
Family: Physeteridae
Total length: maximum 13 feet (4 m)
Weight: up to 700 lb (320 kg)
Diet: cephalopods, crustaceans, fishes
Gestation: probably about 10 months

Comparatively short head; 9–12 teeth on each side of lower jaw. Dorsal fin. Black back, grey belly.

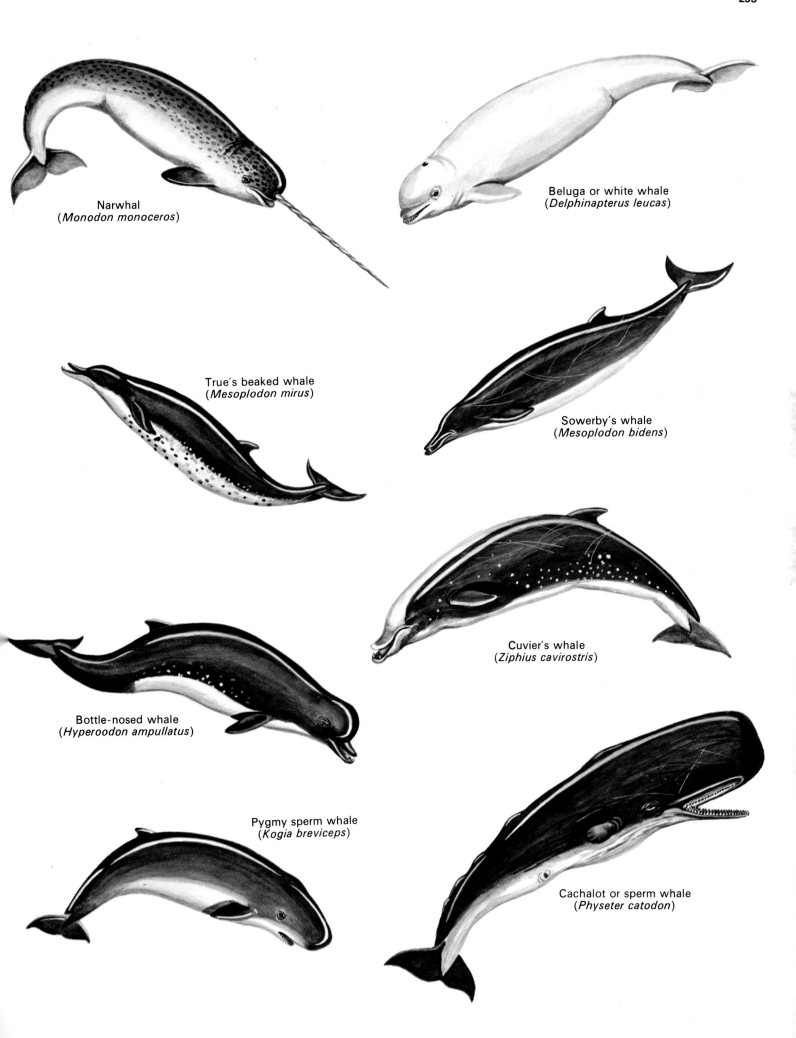

The sperm whales

Apart from the huge whalebone whales the species that has been hunted most intensively over the centuries has been the cachalot or sperm whale. Its economic importance is founded principally on its yield of sperm oil and spermaceti, as well as the ivory from the teeth, part of the skin and the ambergris forming in the intestine.

There is marked sexual dimorphism in this species, males measuring up to about 60 feet, females being much smaller. A characteristic feature is the size of the head which takes up more than one-third of the total length of the body and is strangely shaped, as if topped by a large cushion. In fact the skin does not fit tightly over the skull (as in the case of other vertebrates) but takes the form of a huge pouch or receptacle containing an oily, transparent fluid. This liquid is known as spermaceti but has nothing in common with seminal fluid, for which it was once mistaken. When the liquid comes into contact with air, as is the case when removed from a dead whale, it solidifies into a white wax known as spermaceti, used in the manufacture of candles, lubricants and cosmetics. Unlike the whale oil yielded by whalebone whales, the oil from the blubber of sperm whales is used exclusively for industrial purposes.

In the sperm whale only the lower jaw, small in relation to the bulk of the head, contains triangular, undifferentiated teeth, the number varying from one individual to another but ranging from twenty to thirty on either side. The ivory is much prized by Japanese craftsmen for sculpting figurines and ornaments. The strongest sections of skin, including that of the head, are used for tanning.

The intestine of the sperm whale, which is extraordinarily long (as much as 500 feet), often secretes a greyish or blackish substance, disagreeable to smell when fresh, known as ambergris. This was long regarded (and still is, in some coastal communities) as an aphrodisiac, but is mainly used nowadays for manufacturing perfumes.

It is not known why some sperm whales (but not all) secrete ambergris. Certainly they can expel it, for lumps of the substance are sometimes found floating on the surface or thrown up on beaches. The formation and function are still unknown. Individual lumps may be enormous, perhaps weighing half a ton; and considering the high price paid (at one time £5 per ounce) each represents a small fortune. Although the sperm whale is listed by the IUCN as an endangered mammal, the search for ambergris has had no influence on its numbers, since it is so rarely found actually inside the whale itself.

Mighty undersea hunter

Sperm whales and other toothed species have no baleen plates and thus cannot filter water and feed on the tiny crustaceans eaten by the whalebone whales. For this reason they hunt much larger prey; but although there have been rumours of sperm whales swallowing humans, no evidence has ever been found.

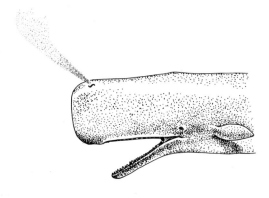

The sperm whale has a single blow-hole on the left side of its head and thus emits a distinctive, oblique-angled spout which is easily recognised by experienced seamen.

The main food items of sperm whales are cephalopods, including octopuses and giant squids. In the quest of such victims the whales are often engaged in pitched battles, as is clearly proved by the gaping wounds and scars which have been revealed on their carcases. Although sometimes the whale has to be content with a tentacle, it often gains an outright victory in such struggles. Thus Robert Clarke discovered a squid weighing over 400 lb, with tentacles 30 feet long, in the stomach of one sperm whale caught off the Azores. The intestines of other whales have also been found to contain the undigested horny jaws of even larger cephalopods. In general, however, the size of prey ranges from 3-6 feet, with up to one hundred of such animals making up a single meal.

Capturing the larger cephalopods demands considerable prowess on the part of the sperm whale; and only an animal so massive and, at the same time, capable of diving to immense depths could possibly achieve such a feat with any degree of regularity. The sperm whale, however, is perfectly adapted to conditions in a marine habitat where there are almost no competitors or rivals; and it has special ways of grasping and killing these slippery monsters of the deep which represent such a valuable source of protein.

The sperm whale plunges down, almost in a nose-dive, to great depths and is capable of remaining under the surface for more than an hour, exploring the surroundings, it jaws agape. Any animal that is unfortunate enough to be caught in that yawning cavity is doomed. As the mouth snaps shut the pointed, peglike teeth lining the lower jaw fit neatly into corresponding

The undifferentiated teeth of the sperm whale are situated on either side of the lower jaw, slotting neatly into alveoli along the edges of the upper jaw. Each tooth weighs one or two pounds.

Over 3,000 feet

Hunting technique of a sperm whale in capturing a giant squid. The slippery prey is gripped firmly between the whale's jaws but the cephalopod will attempt to deploy its tentacles and prevent itself being dragged to the surface.

alveoli or sockets along the upper jaw, and even the most slippery prey is caught firmly as in a trap. Gripping its victim, the whale now shoots straight upwards at high speed. The effect of decompression is often sufficient to kill the cephalopod, but if not, a titanic struggle may ensue as both creatures of the deep employ to the full the vastly different fighting techniques with which they are naturally endowed. The whale endeavours to drag its victim to the surface while the octopus or squid tries to bring its tentacles into play so as to paralyse and drown its predator.

In addition to cephalopods a sperm whale will consume fishes in large quantities. There are written records, for example, of the occasion on a March day in 1874 when the inhabitants of the French port of Audierne panicked at the sight of thousands of fishes leaping up the beaches and clogging the harbour entrance. It was only on the following day that the cause of the phenomenon was discovered when the townsfolk saw thirty-one huge sperm whales struggling against the incoming tide and eventually dying stranded on the beaches. Although most fishes caught by the whale are of moderate size, a 10-foot shark was found in the stomach of one whale caught in 1913 off the coast of Africa.

Harems and bachelor groups

Experienced seamen can identify a sperm whale from a whalebone whale at a distance by its characteristic blow or spout. The former has only one functional blow-hole situated high on the left side of the head, near the front, whereas the latter has a pair of slits positioned centrally and further back. So in the sperm whale the spout consists of a single jet directed obliquely forward instead of vertically. Because of the difference in size it is also a simple matter to distinguish the sex of the sperm whale. This enables zoologists to determine exactly how groups are constituted.

The social behaviour of sperm whales is very elaborate. The animals travel in 'schools' of ten to twenty females and young of all ages, led by an old male. A number of groups may collect together, as is the case along the shores of Patagonia where schools of up to five hundred individuals have been observed. Other males form bachelor groups, while very old bulls, sometimes bearing wounds testifying to their having had to yield supremacy to a younger rival, are often seen on their own.

Females and young evidently prefer warmer waters and seldom stray beyond latitudes 40°N and 40°S, whereas bachelors and older solitary males often visit colder regions, nearing the poles during the summer and returning to the tropics for the breeding season when the ice closes in.

Powerful rutting bulls frequently fight an incumbent dominant male for the possession of a harem, and whalers have described how the rivals seize each other by the jaws, leaping clear out of the water and diving back below the surface in the course of their spectacular battles. Many older males that have been caught bear marks on the lower jaw indicating the

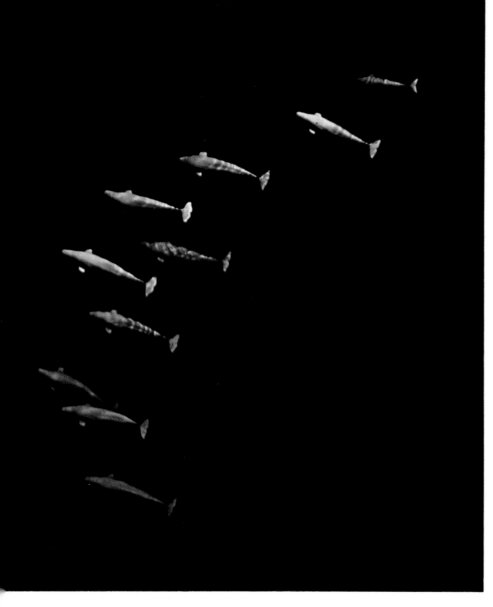

The beluga or white whale, like many of the toothed species, lives in groups or 'schools' numbering up to twelve individuals.

Geographical distribution of the beluga.

places where it has been broken and then knitted together—obviously the consequence of many a battle with a younger and more vigorous competitor.

There are times when a solitary male, toughened by continual contests with rivals and experienced enough to be aware that man is its enemy, will turn really savage and aggressive. Ancient whalers venturing out in open boats would try to avoid meeting such crazed animals which were quite capable of overturning the craft or smashing it to splinters with a violent flick of the tail. Such a creature was Moby Dick, the great white whale, symbol of evil and mystery, so vividly and poetically described by Herman Melville in his famous novel.

Because of contradictory information supplied by various authors little is reliably known of the reproductive habits of sperm whales. Mating seems to take place in a variety of positions, not necessarily at the vertical, as is the case with whalebone whales. But both in the northern and southern hemisphere the animals apparently mate during the warmest season, gravid females remaining in temperate waters until their babies are born. Even the period of gestation is unknown, although most zoologists state that it lasts about sixteen months. Twin births are rare, and the newborn calf measures 10-17 feet, weighing a little more than a ton.

The milk of the female sperm whale, like that of all cetaceans, is extremely nourishing, and the baby grows rapidly. It is probably suckled for twelve or thirteen months.

BELUGA
(*Delphinapterus leucas*)

Class: Mammalia
Order: Cetacea
Family: Monodontidae
Total length: maximum 18 feet (5 m)
Weight: up to about 1,500 lb (675 kg)
Diet: cephalopods, crustaceans, fishes
Gestation: about one year
Number of young: one

Male a little larger than female. Rounded head; trace of beak. Large flippers. From 8–10 teeth on each side of lower jaw. Baby grey or black, later marked with blue or yellow, becoming white when adult at four or five years of age.

Males and females reach sexual maturity at different ages. Although the male may reach this stage when only two years old, the chances are that it will be several years more before he dares to challenge and usurp the place of the leader of the harem. Female sperm whales living in the South Pacific are fit to procreate at the age of about eighteen months but are unlikely to become mothers for some years to come; those inhabiting the North Pacific are not sexually mature until they are about four years old. In the course of adult life female sperm whales will give birth once every two or three years (four years in the opinion of some experts).

Sperm whales appear to take great care of their young and come to the rescue of calves at the first appeals for help. In the bad old days whalers would take advantage of this. They would catch calves and in this way attract the adults. Maternal solicitude extends even further, for a mother will reputedly take hold of a wounded baby and raise it to the surface so that it could breathe.

Amazingly agile for their size, sperm whales often leap high out of the water, coming down with a tremendous thud which can be heard for miles. Sometimes they assume an upside-down position, slapping the water violently with their tails. It is said that the whales are deep sleepers, and that instances of such individuals overturning boats have probably occured when the animals were disturbed while asleep.

Like all Odontoceti, the sperm whale is credited with being a playful and intelligent animal. Hans Hass claims to have heard members of a small school communicating with one another under water, and there are a number of reports of whales having been sighted playing on the surface.

The pygmy sperm whale only remotely resembles its larger relative. It seldom measures more than 12 feet in length and the head occupies only about one-sixth of the body. Although quite common, little is known of its behaviour.

Narwhal and white whale

In the Middle Ages the horn of the so-called unicorn was worth its weight in gold. Although nobody had ever seen the animal to which this singular weapon belonged, the horn attributed to it was spiral in structure, slender, about 6 feet in length, and ended in a blunt tip. The unicorn was popularly described as a strange form of horse with cloven hooves like those of deer, and the horn jutted from its forehead. Its horn was reputed to be an antidote to all known poisons, which encouraged kings and princes to have it crushed and poured into their beverages. Better still if the drinking cup or bowl were itself made of unicorn ivory. Receptacles of this type still exist and the materials have been identified as being of two origins, either from the horn of the Indian rhinoceros or from the single tusk of the marine mammal known as the narwhal.

The narwhal, related to dolphins, measures approximately 17 feet from tip of snout to tail. But the total length exceeds this considerably. The animal has two teeth, both situated

Facing page: The beluga, like all cetaceans, has to rise to the surface from time to time in order to release the air in its lungs. The blow or spout does not consist of a jet of water, as is commonly believed, but of a cloud of mucus vapour in the form of tiny droplets.

The horn of the legendary animal known as the unicorn was thought to possess miraculous medicinal properties. In many cases the horn was actually the tusk of a marine mammal now identified as the narwhal, related to whales and dolphins.

Geographical distribution of the narwhal.

NARWHAL
(*Monodon monoceros*)

Class: Mammalia
Order: Cetacea
Family: Monodontidae
Total length: maximum 17 feet (5 m) excluding tusk
Diet: cephalopods, crustaceans, fishes
Gestation: unknown
Number of young: one

Female and young resemble beluga but in male one of two teeth (the left) is transformed into a spiralled horn, up to 9 feet or even longer. Flippers measure about 12 inches; no dorsal fin. Young bluish-grey; adult lighter, with irregular darker marks.

in the upper jaw. In the case of the adult male one of these teeth grows to an abnormal size, extending well beyond the mouth. Transformed into a stiletto-like tusk, it juts forward to a length of 9 feet or more. No scientist has been able to explain exactly what use the tusk is to its owner. It has been said that it can be used in the event of the narwhal wishing to break a hole in the ice for breathing purposes. Others claim that it may be helpful in tearing up the flesh of squids on which it feeds, for stirring up the ocean bottom to procure prey or for protection against killer whales. Yet if these theories were true the females, who normally do not possess such a tusk, would be incapable of breathing, feeding or defending themselves. More probably it plays some part in social life, and it may have something to do with deciding matters of status, perhaps as an object of sexual selection.

The beluga, also known as the white whale, is the other member of the Monodontidae, measuring 15-18 feet and living in family groups or in small schools of five to twelve individuals. The species is a typical inhabitant of the Arctic and circumpolar regions and only a few isolated animals venture much farther south. One white whale did, however, to the great surprise of zoologists and ordinary people alike, appear in the spring of 1966 in the heavily polluted waters of the Rhine, where it remained for more than a month. The food of this species consists mainly of cephalopods, crustaceans and fishes, and there appears to be a marked preference for sole, plaice and related species inhabiting the sea bed.

BIBLIOGRAPHY

Burton, Robert *The Life and Death of Whales* (Deutsch, New York 1973)
Burton, Robert, Devaney, Carole and Long, Tony *The Living Sea* (Orbis, London 1976; Putnams, New York 1976)
Butler, Jean Campbell *Danger Shark!* (Hale, London 1965)
Carrington, Richard *A Biography of the Sea* (Chatto & Windus, London 1960)
Carson, Rachel L. *The Sea Around Us* (Oxford University Press, London 1961)
Davies, David H. *About Sharks and Shark Attack* (Shuter & Shooter, Pietermaritzburg 1964)
Deacon, G. E. R. *Oceans* (Hamlyn, London 1962)
Douglas, John Scott *The Story of the Oceans* (F. Muller, London 1953; Greenwood Press, Westport CT. 1952)
Fisher, James and Lockley, R. M. *Sea Birds* (Collins, London 1954)
Gunther, Klaus and Deckert, Kurt *Creatures of the Deep Seas* (Allen & Unwin, London 1956)
Herring, Peter J. and Clark, Malcolm R. *Deep Oceans* (Arthur Barker, London 1971)
Idyll, C. P. *Abyss* (Constable, London 1964)
Marshall, N. B. *Aspects of Deep Sea Biology* (Hutchinson, London 1954)
Ommanney, F. D. *The Ocean* (Oxford University Press, London, New York and Toronto 1949)
Phillips, Craig *The Captive Sea* (Chilton Books, Philadelphia and New York 1964)
Russell, F. S. and Yonge, C. M. *The Seas* (Warne, London 1963)

INDEX

Accipitridae: 95
Accipiter poliogaster (grey-bellied goshawk): **142**, 145
Acipenser sturio (common sturgeon): 263, **264**
Acorn barnacles (*Balanus*): 30, 31, **32**, 38, 39
Aethia cristatella (crested auklet): 73
Aipysurus: 65
Albatrosses: Diomedeidae: 268, 269, 271, 276, 277, 277
Albatross, wandering: *Diomedea exulans*: 270–271, 272, 273, 274, 277
Alcidae: 80, 92–95, 97, 144, 199
Alca torda (razorbill): 68–69, 71, 76, 78, 80, 82–83, 92, 95, 95, 102, 276
Amphiprion percula (clownfish): 126, 131, 136, *136*, 137
Amphiura: 162, *162*
Anarhynchus (wrybill): 74, 84
Anarhynchus frontalis (wrybill): 74, 82–83, 84, *84*
Anchovy: *Engraulis enrasicolus*: 171, 217, 214, 215
Anemone fish: *Amphiprion percula*: 126, 131, 136, *136*, 137
Angelfishes: 131
Anglerfish, deep-sea: 16, 25, 168, 178, 191, 193: also *Melanocetus johnsoni*: 238
Anguilla anguilla (common or European freshwater eel): **259–263**, 260, 261, 262
Aptenodytes forsteri (emperor penguin): *140*, *141*, *143*, **143**, 144, 146, **147–149**, 149, 155
Aptenodytes patagonica (king penguin): *141*, 144, **154–155**, 155
Aracari: 125
Architeuthis (giant squid): *239*, 246 247, 295
Arctocephalus: 120; Fur seals: 123
Arenaria interpres (turnstone): 70, 70, 74
Arenicola marina (lugworm): 41, 46, 162, *163*
Argus, blue-spotted: *Cephalopholis argus*: 42–43
Argyropelecus affinis (hatchetfish): *238*, *243*
Asterina gibbosa (cushion star or five-pointed star): 38
Astropecten (sea stars): 162
Astrophyton (basket stars): 133

Atherinidae: 189
Auk, little: *Plautus alle*: 73, 80, 93, 203
Aulostomus maculatus (trumpetfish): 133, *133*, *135*, 136
Avocet: *Recurvirostra avosetta*: 46, 51, 73, 74, 77, **79**, 80, **82–83**

Balaena mysticetus (Greenland right whale): 282, 283, 286, 291
Balaenoptera acutorostrata (lesser rorqual): 282, 284
Balaenoptera borealis (sei whale or Rudolph's rorqual): 282, **283–284**
Balaenoptera brydei (Bryde's whale): 282, 284, 286
Balaenoptera musculus (blue whale): 18, 20, 56, 142, 281, 282, 283, **285–286**, 287–288, 292
Balaenoptera physalus (common rorqual or fin whale): 146, *149*, 282, **283–284**
Barnacles, stalked: 31, 36
Barracuda: *Sphyraena*: 137, **174–175**
Barracuda, Californian: *Sphyraena argentea*: 170
Barracuda, great: *Sphyraena barracuda*: 169
Bass: *Dicentracus labrax*: 31, 43, 102
Bathylagus: 121
Bathylagus callorhinus (no common name): 121
Beluga, giant: *Huso huso*: 263–264
Berardius: 228
Bichir: *Polypterus bichir*: 197
Blenniidae: 36, *40*, 41
Blenny, false cleaner: *Aspidontus taeniatus*: 134, *134*, 137
Bluefish: *Pomatomus saltator*: 169, 174
Bonito, oceanic (or skipjack): *Katsuwonus pelamys*: 265
Boobies: 202–212
Boutu: *Inia geoffrensis*: 285
Boxfish: *Ostracion lentiginosum*: 195, *195*
Bream, sea: *Diplodus*: 31, 43, *45*, 102
Bristle worms: 36, *102*, 137, 162
Brittlestar (see *Amphiura*, *Amphophiura*): 31, 162, 244
Burhinidae: 80, 83
Burrfish, striped: *Chilomycterus schoepfi*: *169*
Butterflyfishes: *131*, 136, 168

Callorhinus: 120, 123
Callorhinus ursinus (Alaska fur seal): 120, 123, *123*
Cancer (crab): 162
Capella: 70, 77
Capella gallinago (snipe): 70, 77
Carcharinidae: 180, *250*, *257*
Carcharodon: 251
Carcharodon carcharias (great white or blue pointer shark): 251
Carcinus: 31–32, 34, 38, 49
Carcinus maenas (shore crab): *31–32*, 34, 38, 49
Cardium (cockle): *41*, 162
Cardium edule (common or edible cockle): 48, *163*
Caretta: 58
Caretta caretta (loggerhead turtle): 52, *53*, **58–59**, *58–60*
Carpet shell, cross-cut: *Venerupis decussata*: 48
Catharacta: 143, 146, *149*, **152–153**, *153*, 277
Cephalopholis: 42–43
Cephalopholis argus (blue-spotted argus): 42–43
Cephalorhynchus: 222, 229
Cephalorhynchus commersoni (Commerson's dolphin): 222, 229
Cerianthus: 32–33
Cetorhinus maximus (basking shark): 251, *255*, 257
Chaenopsis (toadfish): 36
Charadrius: 82, 84
Charadrius apricarius (golden plover): 70, *70*, 72, 80, 84
Chauliodus sloanei (viperfish): *238*
Cheirodon water fishes of family Characidae: 34–35
Cheirodon axelrodi (cardinal tetra): 34–35
Chelmon rostratus (longnose butterflyfish): *131*
Chelonia mydas (green turtle): 52, *53*, 54, *55*, 55, 56, **57–58**
Chionididae: 82, 83, 87, 146, **152**, *153*
Chionis alba (wattled sheathbill): 73, 80, 87, *87*, 148, **152**
Chionis minor (lesser sheathbill): 87, 143, 148, *148*, **152**
Chiton (coat-of-mail shells): 30, 31, 38, *39*, 102
Chlidonias niger (black tern): 70, 73, 75

Chondrichthyes: 178, 197
Ciconia (stork): 201
Cirripedia (barnacles): *102*, 243
Clams: 32, 36, *102*, 162, 132, **134**
Clingfish, shore: *Lepadogaster lepadogaster*: 41
Clingfish, two-spotted: *Diplecogaster bimaculatus*: 41
Clownfish: *Amphiprion percula*: 126, 131, 136, *136*, 137
Clupeidae: 172, 271
Clupea harengus (herring): *169*, 172, *172*
Cockle, common or edible: *Cardium edule*: 48, *163*
Cod: *Gadus morhua*: 178, *196–197*
Coelacanthidae: 197, *236*
Conus (cone shells): 133, 137, *137*
Conger fish (conger eel): *169*, 178, 193, *260*, *261*
Coris angulata (twinspot wrasse): *131*
Cormorants: 77, 203, *203*, 204–205, 206, 214, 215, 277
Crabs: 31–32, 38, 102, 163
Crangon (shrimp): 162
Crenilabrus (wrasses): 43
Crocethia (sanderling): 72, 85, 86
Crocethia alba (sanderling): 72, 85, 86
Cursorius cursor (cream-coloured courser): 97
Cushion star: *Asterina gibbosa*: 38
Cystophora cristata (hooded seal): *107*, 117

Dasyatidae: 184–186, *186*
Delphinapterus (white whale): 228, 278–279, 281, 285, *293*, 297, *297*, 298–299, **300**
Delphinapterus leucas (beluga or white whale): 228, *278–279*, 281, 285, *293*, 297, *297*, 298–299, **300**
Delphinus (common dolphin): 226, 228
Delphinus delphis (common dolphin): 222, 227, **230**
Dentalium (elephant tusk shell): 162, 243
Dermochelydae: 52, *53*
Dermochelys coriacea (leathery turtle or luth): 50, 51, 52, *53*, 61, **62–64**, *62–63*
Devilfish, mobular or manta ray: *Mobula mobular*: 186
Diadema: *135*, 137

Dicentrarchus (bass): *31*, 43, **102**
Dicentrarchus labrax (bass): *31*, 43, *102*
Diodontidae: 194
Diomedeidae: *268–269*, 269, 271, *276*, 277, *277*
Diomedea exulans (wandering albatross): *270–271*, 272, *273*, 274, 277
Diomedea melanophris (black-browed albatross): 274
Diplecogaster bimaculatus (two-spotted clingfish): 41
Diplodus (sea bream): *31*, 43, 45, *102*
Dogfish, lesser spotted: *Scyliorhinus caniculus*: 168, *178–179*, 180, *180*, *197*
Dolphins: 3, *216–217*, *218–219*, 222, 223, *224–225*, *229*, **230–233**, 285
Donax trunculus (wedge shell): 48, *162*
Dromadidae: 83
Dromas (crab plover): 72
Dromas ardeola (crab plover): 72, *80*
Dugong dugong (dugong): 101, 123, *123*

Eagle, African fish: *Haliaeetus vocifer*: 95, *98*
Eagle, Madagascar fish: *Haliaeetus vociferoides*: 95, *98*
Eagle, Pallas' sea: *Haliaeetus leucoryphus*: 95, *98*
Eagle, Sanford's or Solomon Island sea: *Haliaeetus sanfordi*: 95, *98*
Eagle, Steller's sea: *Haliaeetus pelagicus*: 95, *98*
Eagle, white-bellied: *Haliaeetus leucogaster*: 95, *98*
Eagle, white-tailed: *Haliaeetus albicilla*: 70, 77, 78, **95–99**, *98*, *102*
Echinocardium (heart urchin): 162
Echinus: 159, *160*
Echinus melo (sea-urchin): *160*
Eel, common or European freshwater: *Anguilla anguilla*: **259**, 263, *260–262*
Eel, conger: *Conger conger*: 169, 178, 193, *260–261*
Eel, moray: *Muraena helena*: *26–27*, 43, 137, *169*, 178, *192*, 193
Eider: *Somateria mollissima*: 77, *77*
Elephant tusk shell: *Dentalium*: 162, *243*
Emydocephalus: 65
Engraulidae: 214
Engraulis: 214
Engraulis encrasicholus (anchovy): *171*, *207*, 214, *215*
Enhydra lutris (sea otter): 101, **103–106**, *102–106*
Ensis (razor shell): 48
Enteropneusta (acorn worm): 46
Epinephelus: 131
Epinephelus adscensionis (rock cod): 131
Eques: 131
Eques lanceolatus (jackknife fish): *131*
Eretmochelys: 58
Eretmochelys imbricata (hawksbill turtle): 52, *53*, **58**, *58–59*
Eretmophorus: 241
Eretmophorus kleinenbergi: *241*
Erignathus: 117
Erignathus barbatus (bearded seal): 117
Eschrichtiidae: 281
Eschrichtius gibbosus (Californian grey whale): 227, *232*, *282*, 283, *287*
Eubalaena australis (black right whale): 283
Eubalaena glacialis (Biscayan or North Atlantic right whale): *282*, 283, 286, 291
Eubalaena japonica (Japanese whale): 283
Eumetopias jubatus (Steller's sea-lion): 120, *120*

Eunicella (sea fan): 160
Euphausia: 20, 142
Euphausia superba (krill): *18*, 20, 142, *142*, *146*, *149*, 285
Eurynorhynchus: 74
Eurynorhynchus pygmaeus (spoonbill sand piper): 74, *82–83*, 84

Falco eleonorae (Eleonora's falcon): 78, 79, *102*
Falco peregrinus (peregrine falcon): 70, 77, 78, *102*, 248
Falcons: 70, 77, *78–79*, 102, 248
Fan shell: *Pinna squamosa*: 164
Feresa: 229, 284
Feresa attenuata (pygmy killer whale): 229, 284
Fish, cardinal: *Siphamia versicolor*: *135*, 137
Fish, deep-sea: *Eretmophorus kleinenbergi*: 241; *Lamprotoxus flagellibarba* and others: 238
Fish, jackknife: *Eques lanceolatus*: *131*
Flatfishes: 193, *193*
Flounder: *Platichthys flesus*: *2*, 43, 178, *178*, 193
Fratercula arctica (common puffin): 70, 73, 77, 80, 88, 93, 94, *96*
Fregata aquila (Ascension frigate bird): 210
Fregata ariel (lesser frigate bird): *202*, 210
Fregata magnificens (magnificent frigate bird): *202*, 206, *206*, 209
Fregata minor (great frigate bird): *202*, 209–210
Frigate birds: 202, 206, *206*, 209, 210
Frogfishes: 193
Fulmarus glacialis (fulmar): *274*
Fungia, genus of solitary stony coral: 133

Gadiformes: 197
Gadidae: 178, *196–197*
Galeocerdo cuvieri (tiger shark): 255
Gannets: 202, 210, **212–213**, *214*, 215
Gapers: *Mya*: 41, 162
Garfish (*Belone bellone*): *102*
Glareolidae: 80, 83
Glareola pratincola (pratincole): 70, 73, 74, *80*, *82–83*
Globicephala (pilot whale): 226, 227, 228, 229
Globicephala melaena (caa'ing or pilot whale): *22*, *223*, 228, 229, 284
Glossobalanus (acorn worm), see also *Balanoglossus*: 46
Gobiesocidae: 41
Gobiidae: 16, 41
Godwit, black-tailed: *Limosa limosa*: 72, 87
Goshawk, grey-bellied: *Accipiter poliogaster*: **142**, *145*
Grampus griseus (Risso's dolphin): *222*, 229
Groupers (family Serranidae): 135
Guillemots: *68–69*, 71, 76, 77, 78, 79, 80, 92, 94, 95, 203, 277
Guitarfish: *Rhinobatos rhinobatos*: 168, *176*, 182
Gulls: Laridae: 71, 75, 78, *80*, 88, **89–92**, *102*
Gurnards: Triglidae: *177–178*, 191
Gygis alba (fairy tern): *73*
Gyrfalcon: *Falco rusticolus*: 79

Haddock: *Melanogrammus aeglefinus*: 178
Haematopodidae: 80, 83
Haematopus ostralegus (common oyster-catcher): *82–83*
Hagfishes: *Myxine*: 197
Hake: *Merluccius merluccius*: *168–169*, 178, 189, *189*, 197

Haliaeetus (eagles and sea eagles): 70, 77
Haliaeetus albicilla (white-tailed sea eagle): 70, 77, 78, **95–99**, *98*, *102*
Haliaeetus leucocephalus (bald or American eagle): 95, *98*, 99
Haliaeetus leucogaster (white-bellied sea eagle): 95, *98*
Haliaeetus leucoryphus (Pallas' sea eagle): 95, *98*
Haliaeetus pelagicus (Steller's sea eagle): 95, *98*
Haliaeetus sanfordi (Sandford's or Solomon Island sea eagle): 95, *98*
Haliaeetus (Cuncuma) vocifer (African fish eagle): 95, *98*
Haliaeetus vociferoides (Madagascar fish eagle): 95, *98*
Halibut: *Hippoglossus hippoglossus*: 178
Halichoerus grypus (grey seal): *107*, 109, *112*, *112*, 113–116, *114*, *115*, 117
Halocynthia: *31*, 160
Harrier, swamp: *Circus approximans*: 78
Hatchetfish: *Argyropelecus affinis*: 25, *238*, *243*
Heart urchin, purple: *Spatangus purpureus*: *163*
Heptranchias cinereus (seven-gilled shark): 250
Hermit crabs: *31*, 38, *48*
Herring: *Clupea harengus*: *16*, *166–167*, 169, **172–173**, *172*
Heterodontidae: 180
Hippocampus guttulatus (sea-horse): *31*, *43*, *43*, *102*, 169
Hippoglossus (halibut): 178
Histriophoca (ribbon or banded seal): 117
Holothuria (sea cucumber): *31*, 34, 38, 194, 243
Homarus (lobsters): 162
Horn sharks: Heterodontidae: 180
Huso huso (giant beluga): 263
Hydrobatidae: 277
Hydrophidae: 65
Hydrophinae: 67
Hydroprogne (Caspian tern): 73, 90, 91
Hydroprogne caspia (Caspian tern): 73, 90, 91
Hydrurga (leopard seal): 113, 116–117, 143, 146, 149
Hyperoodon ampullatus (bottle-nosed whale): 285, *293*

Jacanidae: *80*, *82*, 83
Jellyfish: 20

Katsuwonus pelamys (skipjack or oceanic bonito): 265
Kittiwake: *Rissi tridactyla*: 77, 78, 78
Kogia breviceps (pigmy sperm whale): 285, 292, *293*, 298
Krill: *Euphausia superba*: *18*, 20, 142, *142*, *146*, *149*, 285

Labridae: 77, 189
Labroides (cleaner wrasses): 134, *134*, 136
Labrus (cuckoo wrasses): 43, *169*
Lagenodelphis hosei (Bornean dolphin): 229
Lagenorhynchus: *216*, 227, 228, 229
Lagenorhynchus acutus (white-sided dolphin): 222
Lamna nasus (porbeagle or mackerel shark): 250
Lampetra fluviatilis (lamprey or lampern): 197
Lantern shells: *Thracia*: 162
Lapwings: Charadrii: *80*, 84
Larus argentatus (herring gull): 70, 75, *80*, *82*, 83, 89, 92, *92*, 277
Larus pipixcan (Franklin's gull): 88
Laticauda: 65, 66

Laticauda semifasciata (sea snake): 67
Laticauda colubrina (flat-tailed sea-snake): 66–67
Latimeria chalumnae (coelacanth): 197, *236*, *237*, 243
Leander (prawn): 162
Lepadogaster lepadogaster (shore clingfish): 41
Lepas (ship's or stalked barnacles): 36
Lepidochelys kempi (Kemp's Ridley turtle): 52, **60–62**, *60*
Leptonychotes weddelli (Weddell seal): *17*, 109, 116, *116*, 142, *146*, *149*
Lily-trotter: Jacanidae: *80*, *82*, 83–85, *84*, *85*
Limosa (godwit): 70, 72
Limosa limosa (black-tailed godwit): 72, 87
Limpet: *Patella*: *31*, 32, 36, *38*, 39, *102*
Lionfish: *Pterois volitans*: 137
Lipotes vexillifer (Chinese river dolphin): 228
Lissodelphis borealis (northern right whale dolphin): 222
Lithotryta (acorn shells): 161
Littorina (periwinkles): 37, 102
Lobsters: *156*, 157, 162, 163, 164
Loligo: 244
Loligo vulgaris (common squid): 244
Lophiidae: 193
Lophius piscatorius (anglerfish): *169*
Lotus-bird: *80*, *82*, 83
Lugworm: *Arenicola marina*: 41, 46, 162, *163*
Lunda (tufted puffin): 93
Lungfish, African: *Protopterus annectans*: 197

Mackerel: 169, *173*, *173*, 174
Mackerel, Spanish: *Scomber colias*: 174
Mackerel sharks: Isuridae: 250
Macoma: 162
Macronectes giganteus (giant fulmar or petrel): *146*, 269, *275*, 277, *277*
Maia (spider crab): *31*, 49, 162
Makaira (marlin): 265
Makaira albida (Atlantic white marlin): 265
Manatees: *Trichechus*: 101
Manta, Australian: *Mobula diabolis*: 186
Marlin, Atlantic white: *Makaira albida*: 265
Marthasterias glacialis (olive-green or spiny starfish): 38
Medusa: 20, 240
Megadyptes: 155
Megaptera novaeangliae (humpback whale): *146*, 281, *282*, 284–287, 292
Melanitta nigra (common scoter): 70, 77, 277
Melanocetus: 238
Melanocetus johnsoni (deep-sea anglerfish): *238*
Melanogrammus aeglefinus (haddock): 178
Merluccius merluccius (hake): 168, *169*, 178, 189, *189*, 197
Mesoplodon bidens (Sowerby's whale): *293*
Mesoplodon mirus (True's beaked whale): *293*
Microcephalophis gracilis (sea-snake): 67
Millepora (stinging coral): 137
Mirounga (sea elephants): 109, 116
Mirounga angustirostris (northern sea elephant): 109, *109*, 116, 117
Mobula (manta rays or devilfish): 186
Mobula diabolis (Australian manta): 186
Mobula mobular (mobular, devilfish or manta ray): *176*, 186, *187*
Monachus (monk seals): 118, 122

Minachus schauinslandi (Hawaiian monk seal): 118, 122
Monachus tropicalis (Caribbean monk seal): 117, 122
Monkfish: *Squatina squatina*: *168*, *176*, 181, *181*, 182
Monodon (narwhal): 298–300
Moorish idol: *Zanclus cornutus*: 135–136
Mudskipper: *Boleophthalmus*: 206, 207, 208, *209*
Mullidae: *190*, 191, *191*, **192**, *192*
Mullus surmuletus (red mullet): *169*, *190*, 191, 192, *192*
Muraena (moray): 43, 178, 192
Muraena helena (common moray): 26, 27, 43, 137, *169*, 178, *192*, 193
Murex brandaris (rock whelk): *163*
Mussel: *Mytilus edulis*: *31*, 32, *102*
Mya (gapers): *41*, 162
Myliobatidae: 184
Myliobatis aquila (eagle ray): 184, *185*, 186
Mytilus: 32, 102
Mytilus edulis (edible mussel): *31*, 32, *102*

Narwhal: *Monodon monoceros*: 228, 285, *293*, **298–300**, *300*
Natica (necklace shells): 161
Naucrates ductor (pilotfish): 126
Negaprion brevirostris (lemon shark): 250
Nematocarcinus: 243
Neobalaena (pygmy right whale): 283, 286
Neobalaena marginata (pygmy right whale): *282*, 283, 286
Neomerus: 228
Neomeris phocaenoides (finless black porpoise): 230
Neophoca: 120
Neophoca cinerea (Australian sea lion): 120, *120*
Neopilina: 243
Nephthys: 162
Nucula (nut shells or nut clams): 162
Nudibranchia (nudibranchs or sea-slugs): *161*
Numenius arquata (curlew): *70*, 80, 83

Octopus vulgaris (European common octopus): 31, 41, 245
Odobenidae: 114
Odobenus rosmarus divergens (Pacific walrus): 118
Odobenus rosmarus rosmarus (Atlantic walrus): *100*, *101*, 101, 106, *107*, *109–111*, 118, *118*, 119, *119*
Odontoceti (toothed whales): *223*, 227, 280, 281
Ommatophoca rossi (Ross seal): 116, 142, *146*, 149
Orcaella brevirostris (Irrawaddy dolphin): 222, 229
Orchestia (sandhoppers): 37
Orcinus (killer whale): 233–235
Orcinus orca (killer whale or grampus): *17*, *146*, *149*, *223*, 227, 229, **233–235**, *234*, *235*, 284
Orectolobus maculatus (carpet shark): *168*, *176*, 181, *183*
Ostraciontidae: 195
Otariidae: 106, *114*, 119–121
Otaria byronia (South American sea-lion): 120, *120*
Otocyon: 263
Otter, sea: *Enhydra lutris*: 101, **103–106**, *102–106*
Oystercatchers: *70*, *70*, 74, 77, *80*, *81*, *83*, *84*, *85*

Pachygrapsus: 49
Pachyptila (prions): 271
Pachyptila desolata (Antarctic prion): 277
Pagophilus: 117

Pagophilus groenlandicus (harp or Greenland seal): 117
Paguridae: 38
Palaemonidae: *40*
Palinurus vulgaris (spiny lobster or crawfish): *156*, 162, *163*
Paracanthurus hepatus (surgeonfish): *131*
Paracentrotus: 38
Paracentrotus lividus (sea urchin): 38, 159, *160*
Parrot-fish (*Scarus*): *133*
Patella: *31*, 32, 36, *38*, *39*, *102*
Pearlfishes: Carapidae: 194, *194*
Pelecanoididae: 203, 269, 271, *276*, 277, *277*
Pelecanoides (diving petrel): 203, 271, *276*, 277
Pelecanoides garnotii (Peruvian diving petrel): *214*, 215
Pelecanoides magellani (Magellan's diving petrel): *274*
Pelecanoides urinatrix (common diving petrel): *274*
Pelecanus occidentalis (brown pelican): *214*, 215, *277*
Penaeidae: 162
Penguins: *138*, *139*, **143–149**, *143*, *150*, *151*, **155**, 203, *276*, 277
Pennatula (sea pens): 162
Periwinkles: *31*, 37
Petrels: 78, 201, 203, 277
Petricola pholadiformis (American piddock): 32
Petromyzoniformes: 197
Phaethontidae: 202, 207
Phaethon: 209
Phaethon aetherius (red-billed tropicbird): *202*, *209*, 210, *210*
Phaethon lepturus (white-tailed tropicbird): *202*, 210
Phaethon rubricauda (red-tailed tropicbird): *202*, 210
Phalacrocorax: 203
Phalacrocorax aristotelis (shag or crested cormorant): 76, 77, 78, 200, *204*, 205, 205
Phalacrocorax bougainvillei (guanay or Peruvian cormorant): *205*, *214*, 215, *277*
Phalacrocorax capensis (Cape cormorant): *214*, 215
Phalacrocorax carbo (common cormorant): 77, 203, *203–205*, 205, 206, *277*
Phalacrocorax pygmaeus (pygmy cormorant): 203, 206
Phalaropus fulicarius (grey phalarope): 80, 82, 83
Phalaropus tricolor (Wilson's phalarope): 87
Phegornis mitchellii (Mitchell plover): 84
Philomachus (ruffs): 85, 86
Phoca vitulina (common seal): 117
Phocaena phocaena (common porpoise): *223*, 230
Phocaenoides: 228, 230
Phocaenoides dalli (Dall's porpoise): *223*
Phocarctos hookeri (Hooker's or New Zealand sea-lion): 120, *120*
Phoebetria (sooty albatross): 274
Pholadidae: 36
Physalia physalis (Portuguese man-o'-war): 22
Physeter catodon (sperm whale): 228, 281, 284, 285, 291, 292, *293–296*, **294–298**
Picasso fish: *Rhinecanthus aculeatus*: 126
Piddocks: *31*, 36
Piddock, American: *Petricola pholadiformis*: 32
Pilot fishes (*Naucrates ductor*): 126
Pinna (fan shells): 164
Pinnipedia: 55, 101, 103, 106, **106–123**, 279, *279*
Pinnotheres (pea crabs): 38

Pipefishes: *Syngnathus*: 43, 193, 194, *194*
Platichthys flesus (flounder): 43, 178, *178*, 193
Plautus alle (little auk or dovekie): *73*, **80**, 93, 203
Pleuronectiformes: 178, 193, *193*
Pleuronectidae: 168
Pleuronectes (plaice): *168*
Pleuronectes platessa (plaice): *168*, *178*, *178*, 193
Plovers: 70, 72, 74, 77, 80, 82, 83, **84**
Pluvialis: 84
Pluvialis squatarola (grey plover): 84
Polychaeta: 36, *46*, *102*, 137, 162
Polyplacophora: 38
Pomacanthus (angelfishes or butterflyfishes): *131*
Pomacanthus imperator (imperial angelfish): *131*
Pomatomus (bluefish): *169*, 174
Pomatomus saltator (bluefish): *169*, 174
Porbeagle or mackerel shark: *Lamna nasus*: 250
Porcellana: 31
Porcellana platycheles (porcelain crab): 31
Porpoises: 101, *223*, 228, 230
Portuguese man-o'-war: *Physalia physalis*: 22
Portunidae: *31*, 49
Portunus (swimming crabs): *31*, 49
Portunus pelagicus (swimming crab): *31*, 49
Pratincole: *Glareola pratincola*: 70
Prawn: 243
Prion, Antarctic: *Pachyptila desolata*: 277
Prionace: 250
Prionace glauca (great blue shark): 250
Pristidae: 187
Pristis pristis (sawfish): *176*
Pristiophoridae: 187
Procellariiformes: 71, 203, 269–277, *270–274*
Procellariidae: 277
Psammechinus: 38
Pseudorca crassidens (false killer whale): *223*, 229, *234*, 284
Pterodroma (gadfly petrels): 271
Pterois (lionfish): 137, *137*
Pterois volitans (lionfish): 137
Puffins: 71, 76, 77, 78, *80*, 88, 92, 93, 94, *96*, 200, 203, *277*
Puffinus gravis (greater shearwater): 277
Puffinus griseus (sooty shearwater): *274*
Puffinus puffinus (Manx shearwater): 200
Puffinus tenuirostris (short-tailed sheerwater): 274, 276
Pusa: 117
Pygoscelis: 155
Pygoscelis adeliae (Adélie penguin): *141*, 143, *144*, 144, *145*, 146–147, *146*, *149*, 155, 200, 201
Pygoscelis antarctica (chinstrap penguin): *141*, 144, 155
Pygoscelis papua (gentoo penguin): *141*, 155

Rajiformes: 181–184
Raja (skates and rays): 184
Raja batis (common skate): 184, *184*
Raja clavata (thornback ray): 184, *184*, 185
Ratfish: Chimaera: 197
Rays: 32, 43, *44*, *168*, 178, 182, *182*, 184, *185*, 197
Razorbill: *Alca torda*: *68*, *69*, 71, 76, 78, *80*, *82*, *83*, 92, 95, *95*, 102, *276*
Razor shell: 48
Recurvirostridae: *80*, 83
Recurvirostra (avocets): 46, 74, 79
Recurvirostra avosetta (avocet): 46, *51*, *73*, 74, 77, **79**, *80*, *82*, *83*

Redshank: *Tringa totanus*: *70*, 72, 74, 85
Remora: Echeneidae: *187*
Reptantia: 162
Rheidae: 160
Rhincodontidae: 251, *252*, 257
Rhinecanthus aculeatus (Picasso fish): 126
Rhinobatidae: 182
Rhinobatos: 182
Rhinobatos rhinobatos (guitarfish): *168*, *176*, 182
Rissa tridactyla (kittiwake): *77*, 78, *78*
Rorquals: 146, *149*, 252, *279*, 281, *282*, 283, 284, 286, 287, *288*
Rostratulidae: *80*, 83
Rostratula: 80
Rostratula benghalensis (painted snipe): 80
Ruff: *Philomachus pugnax*: *73*, 85, 86
Rynchops nigra (black skimmer): *73*, 76, *80*, *82*, *83*

Sabella (tube worms): 46
Sagitta (arrow-worms): *163*, *164*
Sagitta elegans: 163; *S. setosa*: *164*
Sailfish: *Istiophorus americanus*: 265
Sanderling: *Crocethia alba*: *72*, 85, 86
Sandhoppers: *Orchestia*: 32, 37, *45*; *Talitrus*: 37
Sandpiper: 70, *70*, 85, 86
Sandpiper, spoonbill: *Eurynorhynchus pygmaeus*: 74, *82*, *83*, *84*
Sawfish: *Pristis pristis*: *176*, 187
Scalpellum: 243
Scaphopoda: 162, 243
Sciaenidae: 189
Scolopacidae: 83, 85
Scolopax rusticola (woodcock): *70*, 72, 74, 77, 80, *82*, 83, 85, 86, 296
Scombridae: 173
Scomber colias (Spanish mackerel): 174
Scomber scombrus (common mackerel): *169*, **173**, *173*
Scophthalmus maximus (turbot): 193
Scorpaenidae: 137
Scoter, common: *Melanitta nigra*: *70*, 77, *77*
Scyliorhinus: 180
Scyliorhinus caniculus (lesser spotted dogfish): *168*, *178*, *179*, 180, *180*, 197
Scyllarus (squat lobster): 162
Sea anemone: *30*, *31*, 32, *32*, *33*, 36, 38, *40*, *48*, *102*, *136*, 137, 242
Sea-bear see Alaska fur seal: 120, 123, *123*
Sea-cow: Sirenia: 123
Sea-cucumbers: *31*, *34*, 38, 130, 160, 243, 244
Sea-eagle: *Haliaeetus*: *70*, 77
Sea elephant, northern: *Mirounga angustirostris*: 109, *109*, 116, 117
Sea fan: *Eunicella*: 160
Sea hare: *Aplysia*: *31*, *102*, 159
Sea horse: *Hippocampus guttulatus*: *31*, *43*, 43, *102*, 169
Sea lily: Crinoidea: 133, 242
Sea-lions: 101, 106, 113, 119, *120*, *121*, 280
Sea pen: *Virgularia*: 162, 242
Sea slugs: Nudibranchia: *161*
Sea snakes: Hydrophidae: **64–67**, *64*, 65
Sea squirt: *Halocynthia*: *31*, *160*
Sea stars: *Astropecten*: 162
Sea swallows see terns: 71, *71*, 75, 76, *80*, 89, 92, *102*, *277*
Sea urchins: *Echinus, Diadema, Psammechinus*: *31*, 36, 37, 38, 49, *102*, *135*, 137, 159, *160*, 162, 243
Seals: *17*, 101, *108*, *109*, 117, **119–121**, *142*, **143**, 146, *149*, 292
Seedsnipe, least: *Thinocorus rumicivorus*: *73*, *80*, *82*, *83*

Serpula (tube-worm): 46, *46*
Serranidae: 178, 189
Serranus, genus of fishes of family Serranidae: 178, 189
Serranus guaza (comber or grouper): *169*, 178, 184
Shag: *Phalacrocorax aristotelis*: 76, 77, 78, 200, *204*, *205*, 205
Sharks: *17*, 22, 137, **175**, *176*, 177, 181–182, *185*, 197, **249–257**
Shearwaters: 71, 77, 78, 201, 269 277
Shearwater, greater: *Puffinus gravis*: 277
Shearwater, Manx: *Puffinus puffinus*: 200
Shearwater, short-tailed: *Puffinus tenuirostris*: 274, 276
Shearwater, sooty: *Puffinus griseus*:
Sheathbills: 73, *80*, *82*, *83*, 87, 146, **152**, 153, *277*
Shell, tusk: *Dentalium*: 162, 243
Shipworm: *Teredo*: 161, 162
Shrimps: 31, 36, *40*, 46–47, 162
Silversides, fishes: 189
Siphamia: 137
Siphamia versicolor (cardinal fish): *135*, 137
Sirenia: 101, 103, **123**, *123*
Skate, common: *Raja batis*: 184, *184*
Skimmer: *73*, 76, *80*, *82*, *83*, 89
Skuas: *70*, 73, 78, *80*, *82*, *83*, 89, *102*, 143, *146*, *149*, **152–153**, 277
Smelt, sand: *Atherina presbyter* (family Atherinidae): 189
Snakes: 66, **67**
Snipes: 70, 77, 80
Solea (sole): 192
Solea solea (Dover sole): *169*, 192, *192*
Solen: 48
Solen magtinatus (grooved razor shell): 48
Somniosus microcephalus (Greenland shark or sleeper): *255*
Sotalia guianensis (Guyana white dolphin): 223
Sousa: 228, 229
Sousa teuszi (Cameroon or West African white dolphin): *223*
Sparrowhawk, bi-coloured: *Accipiter bicolor*: *140*, **142**
Spatangus (heart urchins): 162
Spatangus purpureus (purple heart urchin): 162
Sphenisciformes: **155**
Spheniscidae: 155
Spheniscus: **155**
Spheniscus demersus (jackass or black-footed penguin): *141*, 155, *214*, 215
Spheniscus humboldti (Humboldt or Peruvian penguin): *141*, 155, 201, *214*, 215
Spheniscus magellanicus (Magellan penguin): *141*, 155
Spheniscus mendiculus (Galapagos penguin): *141*, 143, 155
Sphyraena (barracuda): 137, 174, 175, *175*
Sphyraena barracuda (great barracuda): *169*
Sphyrna zygaena (hammerhead shark: **175**, **176**, 176, *177*, *255*
Spirographis (fanworm): 46
Sponge, boring: *Cliona*: 161
Sponges, glass: Hexactinellida: 243
Squatina squatina (angel fish or monkfish): *168*, *176*, 181–182, *181*
Squid, common: *Loligo vulgaris*: 244
Squid, giant: *Architeuthis*: 146, *239*, 246, 247, *295*, *296*
Squilla mantis (mantis shrimp): 31, *46*, 47, 49
Starfish: 31, *37*, 162
Starfish, crown-of-thorns: *Acanthaster planci*: *132*, 133
Starfish, spiny: *Marthasterias glacialis*: 38, *46*

Stargazer: *Uranoscopus scaber*: *169*, 189
Stenella: 228
Stenella plagiodon (spotten dolphin): *222*, 228, *228*
Steno bredanensis (rough-toothed dolphin): *223*
Stercorariidae: 80, 89
Stercorarius longicaudus (long-tailed skua): *73*, *82*, *83*
Stercorarius pomarinus (pomarine skua): *277*
Sterna (terns): 92
Sterna hirundo (common tern): *71*, *73*, 92
Sterna paradisaea (Arctic tern): *82*, *83*
Stilts: 74
Stingray, common: *Dasyatis pastinaca*: *176*, 186
Stone curlew: *70*, *80*, *83*
Stonefish: *Synanceja verrucosa*: 137
Strongylocentrotus: 159
Sturgeon: *Acipenser sturio*: *263*, **264**
Suckers (clingfishes): 41
Sulidae: 210
Sula bassana (northern gannet): *202*, 210, *211*, 212, *212*
Sula capensis (Cape gannet): 202, 210, *212*, *213*, *214*, 215
Sula dactylatra (masked booby): *202*
Sula leucogaster (brown booby): *202*
Sula nebouxii (blue-footed booby): *202*
Sula serrator (Australian gannet): 210
Sula sula (red-footed booby): *202*, *211*
Sula variegata (Peruvian booby): *202*, 215
Surgeonfish: *Paracanthurus hepatus*: *131*
Swordfish: *Xiphias gladius*: 265
Synanceja verrucosa (stonefish): 137
Syndosimya: 162, *162*
Syngnathidae: 194
Syngnathus (pipefishes): 31, 43, *102*, 193, 194, *194*

Talitrus (sandhoppers): 37
Tasmacetus: 285
Tasmacetus shepherdi (Oliver's beaked whale): 285
Tellinidae: 162, 163
Tellina radiata (tellin mollusc): *163*
Tellins, Baltic: *Macoma*: 162
Terebellidae: 41
Terebella (marine bristleworms): 162
Teredo (shipworm): 161, 162
Terns: 70, *70*, 71, *71*, 75, 76, *80*, *82*, 83, 89, 90, 91, 92, *102*, *277*
Tetra, cardinal (fish): 34, 35
Tetraodontidae: 194
Thinocoridae: *80*, *82*, *83*
Thinocorus rumicivorus (least seedsnipe): *73*, *80*, *82*, *83*
Thracia (lantern shells): 162
Thunnus (tunny or tuna): 264–267
Thunnus alalunga (long-fin tunny): *265*, 267
Thunnus albacares (albacore): *265*
Thunnus thynnus (tunny or bluefin tuna): **264–267**, *265–267*
Toadfishes: *Chaenopsis*, *Thalassophryne*: 36
Torpedinidae: 182
Torpedo: 32, 43, 182
Torpedo marmorata (marbled electric ray): 32, 43, *44*, 182, *182*, *183*
Torpedo torpedo (torpedo or electric ray): *176*
Tower shells: *Turritella*: 162
Trichechus: 101
Tridacna gigas (giant clam): *132*, **134**
Triglidae: 177
Tringa totanus (redshank): *70*, 72, 74, 85
Tropicbird, red-billed: *Phaeton aethereus*: *202*, *209*, 210, *210*

Tropicbird, red-tailed: *Phaeton rubicauda*: *202*, 210
Tropicbird, white-tailed: *Phaeton lepturus*: *202*, 210
Trumpetfish: *Aulostomus maculatus*: 133, *133*, *135*, 136
Tryblidioidea: 243
Tube worms: 31, *32*, *35*, 46, *46*, *102*, 160
Tuna, yellowfin: *Thunnus albacares*: 265
Tunny or bluefin tuna: *Thunnus Thynnus*: **264–267**, *265–267*
Tunny, longfin: *Thunnus alalunga*: 265, 267
Turbot: *Scophthalmus maximus*: 193
Turnstone: *Arenaria interpres*: *70*, 70
Turritella (tower shell): 162
Tursiops (bottle-nosed dolphin): 227, 228
Tursiops truncatus (bottle-nosed dolphin): 3, *216*, *217*, *219*, *222*, *224*, *225*, 229, **230–233**
Turtles, marine: **52–64**, *60*, *61*
Turtle, green: *Chelonia mydas*: 52, *53–56*, 55, **57–58**
Turtle, hawksbill: *Eretmochelys imbricata*: 52, *53*, **58**, *58*, *59*
Turtle, Kemp's Ridley: *Lepidochelyskempi*: 52, **60–62**, *60*
Turtle, leathery (or luth): *Dermochelys coriacea*: *50*, *51*, 52, *53*, 61, **62–64**, *62–63*
Tusk shell (*Cadulus*): 243

Uranoscopidae: *194*
Uranoscopus (stargazer): *194*
Uranoscopus scaber (stargazer): *169*, 189
Uria aalge (common guillemot): *73*
Uria lomvia (Brünnich's guillemot): *94*

Venerupis: 48
Venerupis decussata (cross-cut carpet shell): 48
Venus: 48, 162, *162*
Venus gallina (striped venus): 48
Venus verrucosa (warty venus): 48
Viperfish: *Chauliodus sloanei*: 238
Virgularia (sea pens): 162

Waders: 81
Walrus: *Odobenus rosemarus*: *100*, 101, *101*, 106, *107*, *109–111*, *118*, 119, *119*
Water flea: *Daphnia*: 44, 45
Waterfowl: Anseriformes: 71
Wedge shell: *Donax trunculus*: 48, 162
Whales: 101, 142, **279–300**
Whales: 101, 142, **279–300**
Whale, Baird's beaked: *Berardius bairdi*: 285
Whale, Biscayan or North Atlantic right: *Eubalaena glacialis*: *282*, 283, 286, 291
Whale, black right: *Eubalaena australis*: 283
Whale, blue: *Balaenoptera musculus*: *18*, *20*, 56, 142, 281, *282*, 283, 285, 286, 287, 288, 292
Whale, bottlenosed: *Hyperoodon ampullatus*: 285, *293*
Whale, Bryde's: *Balaenoptera brydei*: *282*, 284, 286
Whale, Californian grey: *Eschrichtius gibbosus*: 227, *232*, *282*, 283, *287*
Whale, Cuvier's beaked: *Ziphius cavirostris*: 284, 285, *293*
Whale, false killer: *Pseudorca crassidens*: *223*, 229, *234*, 284
Whale, fin see rorqual: 252, 281, 283, 284, 286, *287*, *288*
Whale, Gervais' beaked: *Mesoplodon gervaisi*: 285
Whale, Greenland right: *Balaena mysticetus*: *282*, 283, 286, 291

Whale, humpback: *Megaptera novaeangliae*: *146*, 281, *282*, 284–287, 292
Whale, Japanese: *Eubalaena japonica*: 283
Whale, killer: *Orcinus orca*: *17*, *146*, *149*, *223*, 227, 229, **233–235**, *234*, *235*
Whale, Oliver's beaked: *Tasmacetus shepherdi*: 285
Whale, pilot or caa'ing: *Globicephala melaena*: 22, *223*, *228*, 229, 284
Whale, pygmy killer: *Feresa attenuata*: 229, 284
Whale, pygmy right: *Neobalaena marginata*: *282*, 283, 286
Whale, pygmy sperm: *Kogia breviceps*: 285, 292, *293*, 298
Whale, Sowerby's: *Mesoplodon bidens*: *293*
Whale, sperm: *Physeter catodon*: *228*, 281, 284, 285, 291, 292, *293–296*, **294–298**
Whale, True's beaked: *Mesoplodon mirus*: *293*
Whale, white: *Delphinapterus leucas*: 228, *278*, *279*, 281, 285, *293*, 297, *297–299*, **300**
Whalebone whales (Mysticoceti): *18*, 23, 280, 281–284, *283*, 286–292, *287–290*, 296
Whelk, rock: *Murex brandaris*: *163*
Woodcock: *Scolopax rusticola*: 70, 72, 74, 77, *80*, *82*, *83*, 83, 85, 86, 296
Worms: Annelida: 71
Worms, bristle: *Nephthys*; *Terrabella*: 162
Worms, peacock or fan: Sabellidae: 41
Wrasses: 31, *102*, 189
Wrasse, cleaner: *Labroides dimidiatus*: *134*, *134*, 136
Wrasse, cuckoo: *Labrus bimaculatus*: *169*
Wrasse, twinspot: *Coris angulata*: *131*
Wrybill or wrybill plover: *Anarhynchus frontalis*: 74, *82*, *83*, 84, *84*

Xiphiidae: *265*
Xiphias gladius (swordfish): *265*

Zalophus californianus (Californian sealion): 120, *120*
Zanclus (Moorish idol): *135*, 136
Ziphiidae: 284